W9-ADW-331

NORMAN F. BOURKE
MEMORIAL LIBRARY
CAYUGA CO. COMMUNITY
COLLEGE
AUBURN, N. Y. 13021

Hands

By the same author

Old World Monkeys (Academic Press)
Bigfoot (Cape)
The Roots of Mankind (Allen & Unwin)

and with Prue Napier
A Handbook of Living Primates (Academic Press)

Hands

by

John Napier

Visiting Professor in Primate Biology,
Birkbeck College,
University of London

Pantheon Books, New York

Copyright © 1980 by John Russell Napier

All rights reserved under International and Pan-American Copyright Conventions. Published in the United States by Pantheon
Books, a division of Random House, Inc., New York. Originally
published in Great Britain by George Allen & Unwin, Ltd.

Library of Congress Cataloging in Publication Data

Napier, John Russell.
  Hands.

  Bibliography: p.
  Includes index.
  1. Hand.   I. Title.
QM548.N35   611′.97   79-1976
ISBN 0-394-50783-5

Manufactured in the United States of America

FIRST AMERICAN EDITION

Acknowledgements

Anyone who writes a book about science or indeed about any subject that is rich in facts, figures and ideas is from the beginning up to his neck in debt. He must beg, borrow and steal left, right and centre. His pilfering is usually made respectable by the inclusion of a bibliography, but formal citations do not tell more than a fraction of the story of an author's obligation to colleagues living and dead. Much of one's information is derived from memories of conversations heard, snippets of fact retained and from bright ideas once read, but the source promptly forgotten. To all such anonymous creditors I am grateful and I apologise if their feelings are hurt at finding themselves unacknowledged.

I am happy to take this opportunity of discharging with gratitude some of the debts that have accrued during the writing of this book: to John Barron, John Berry, Audrey Besterman (for most of the diagrams), Michael Bush, Dr and Mrs Gardner (for their generosity in allowing me to use a photograph of Washoe never previously published), Ted Grand, Geoffrey Ainsworth Harrison, the late Sir Gerald Kelly, Martin Leadbetter, Gil Manley, Jonathan Musgrave, Prue Napier, Freda Newcombe, Barry Pike (for help with photographs) and Phillip Tobias. Naturally, the opinions expressed in this book do not necessarily reflect the opinions of these experts.

I am grateful to Williams & Wilkins Company, Inc. of Baltimore for permission to quote an extract from *A Method of Anatomy* by J. C. Grant: to Oxford University Press (associated with Geoffrey Cumberlege) for permission to publish extracts from *Elizabethan Acting* by B. L. Joseph, 1951: to Phaidon Press for permission to use diagrams and photographs from my own articles in *Quest for Man*: to Carolina Biological Readers for allowing me to use photographs and drawings from No. 61, *The Human Hand*. I am particularly indebted to the *Illustrated London News* for photographs of the drawings by Neave Parker, the BBC for the photograph of Sir Brian Horrocks, the *Guardian* for that of the Balinese dancer, the Prado for the photograph of the painting of the Court Dwarf by Velasquez and to the other museums and individual people whose names appear in the captions to the illustrations which they kindly supplied.

I have also relied heavily on the writings of the late Professor F. Wood Jones, of Dr Sarah Holt, Mr Michael Barsley and Dr Walter Sorell. My apologies if I have failed to credit any author or source.

87075

Contents

List of Illustrations and Figures

Frontispiece Dürer's "Praying Hands"

List of Illustrations

Hands

Part One

Nature and Evolution of the Hand

Chapter One
"You need hands..."

You need hands

Hands is a book about the anatomy, function and evolution of the hand in man, but since man is not a one-off product of nature but a species with close non-human relatives it is also about the hands of monkeys and apes. Because the functions of the hand touch on so many different aspects of everyday living I have included brief accounts of the role of some of these activities such as tool-making, grooming and gesture. Such important social issues as handedness and the uniqueness of fingerprints are also discussed.

I hope that *Hands* will appeal to a wide audience of scientists and non-scientists who are, one way or another, involved in the hand whether as surgeons or dancing teachers, anatomists or fingerprint officers, palmists or geneticists, or just plain citizens who wish to know more about a fascinating subject. With such a wide audience in view I have tried to write in a light and readable manner, avoiding jargon where possible – or explaining it. Non-scientists may find some parts heavy going but I hope that this will not impair their understanding. The gap between scientists and non-scientists is not so wide as it used to be, largely because of the influence of radio and television and other forms of science reporting. Equally, I hope that scientists will not be offended by a not-too-technical approach. They may even find it a relief, for, as Margaret Mead put it, "to a physicist even a botanist is a layman". We are all laymen once we are outside our expertise. There may be only one language of science but there are many dialects.

I can't say that I learned very much about the hand when I was a medical student: I don't believe that many of us did. Certainly it was not a subject of special teaching, and therefore we learnt little of its function or comparative anatomy and nothing at all of what might be termed "the aesthetics". Our eyes remained firmly closed to the niceties.

It wasn't until much later, after I was qualified, that I became infatuated with the hand, and embarked on a relationship that has lasted for well over thirty years. Another thirty years and I may begin to understand what the hand is really all about. I remember quite vividly the occasion when the penny first dropped, and I discovered that aesthetically the hand was the most beautiful part of the human body. From then on I began to see all around me the beauty of hands in the context of work and play: in the elegant hands of sculptors, artists and musicians, as well as in the rough and powerful hands of the village blacksmith.

1. The powerful hand of David by Michelangelo expresses the position of rest. Compare this figure with Fig. 27.

The hand at rest is beautiful in its tranquillity, but it is infinitely more appealing in the flow of action. The hand of Michelangelo's "David" (Fig. 1) is magnificent but not to my mind so exciting as the hands of Velasquez' court dwarf. David's powerful hand is frozen at rest while the hands of Velasquez's dwarf are frozen in the very act of shuffling a pack of cards. To look at the face as well as the hands is to understand the distinction. David is looking noble but his hand is uninvolved, whereas the hand of the dwarf is very much involved; he is in the process of sliding off the bottom card from the pack. He is no doubt up to one of his conjuring tricks and his sly and arrogant nature is apparent in his evil little face (Fig. 2).

When the hand is at rest, the face is at rest; but a lively hand is the product of a lively mind. The involvement of the hand can be seen in the face which is in itself a sort of mirror to the mind. One of the saddest sights there is is to watch the hands of the mentally disturbed. When the brain is empty, the hands are still.

My own introduction to the hand came from a young plastic surgeon from New Zealand, John Barron, who was giving an informal talk about the hand to a group of house surgeons at the hospital where we all worked. The gestures he used to emphasise his points were so graphic and so eloquent that I became quite hypnotised and he might have been talking in the Maori language for all that I remember of what he said. I recall going on my late-night ward-round after the talk was over, convinced that the hand was going to be my specialisation; and in spite of distractions during my career that led me in different directions, I have always returned to the problems of the hand with a feeling of homecoming, so ubiquitous is it in its involvement in human affairs.

My initial interest in the injuries and functions of the human hand has broadened over the years to include its evolution, comparative anatomy and fossil history as well as the development of tool-use in early man. Additionally I have become involved in such side issues as fingerprinting, use of gesture, handedness, cave art and so on. Most of these intriguing aspects of the subject are included in this book, though not in any great depth.

2. Velasquez's Court Dwarf at the time of Philip IV. The facial expression is one of slyness which leads one to the conclusion that the hands are up to something equally sly. Undoubtedly the arrogant dwarf is performing a conjuring trick (Courtesy of the Prado, Madrid).

You need hands

Strangely enough (and I shall never know whether it was cause or effect), conjuring has always been my principal hobby. No activity demands more of the hands than what is known as "manipulative magic". Unfortunately I was never much good at it, but I learned more about the range and variety of skilled hand movements than at any other time in my life.

All in all, I feel reasonably well qualified to write about some of the aspects of the hand that have interested me. In this sense this book constitutes a personal view and is in no sense an attempt at a definitive study of a very subtle and complicated organ.

There is nothing comparable to the human hand outside nature. We can land men on the moon but, for all our mechanical and electronic wizardry, we cannot reproduce an artificial fore-finger that can feel as well as beckon.

But although our hands are such wondrous organs, we tend to take them for granted. Visitors to the zoo indulge in transports of delight at the way an elephant reaches for an apple with its trunk, and become ecstatic at seeing a squirrel use its paws to eat, but give not a moment's thought to the ineffable capabilities of their own hands.

In fact we treat them abominably – with none of the finesse that we lavish on cameras or hi-fi records. This is true of both sexes, but men are particularly prone to mistreat their hands. Fortunately the hand is adapted against such ill-usage by means of thick skin, horny callosities and a layer of tough fibrous tissue (called the palmar aponeurosis) beneath the skin at the wrist and in the palm of the hand. Even so, infections of the hand are commonplace and injured hands constitute the commonest of all industrial injuries.

The human hand, as well as being the principal vehicle of motor activity, is the chief organ of the fifth sense, touch. With the eye, the hand is our main source of contact with the physical environment. The hand has advantages over the eye because it can observe the environment by means of touch, and having observed it, it can immediately proceed to do something about it. The hand has other great advantages over the eye. It can see round corners and it can see in the dark. The hands are situated at the end of long highly flexible arms which allow the sensory and motor activities to function at some distance from the body; this form of remote control permits one to feel around corners and so solve one of the great problems facing modern man in this technological age: it enables one to adjust the knobs at the back of a television set while keeping one's eyes on the screen! Touch as a function of the hand in appreciating three-dimensional art has never been officially encouraged in art galleries; but in 1976 the Tate Gallery in London put on an exhibition of sculpture especially for the blind and encouraged unlimited handling.

How much more telling it would be for children to use their hands to feel the thickness of the fur or the texture of the skin of a mounted exhibit in a natural history museum than to simply stand and stare. The Smithsonian Institution in Washington, DC, has mounted a large square of elephant skin alongside the main exhibit and they encourage children to handle it as much as they wish.

Dr Leo Harrison Matthews in his book *The Life of Mammals* quotes an example of visual-tactile teamwork. When a friend produces an unusual curio from his pocket you exclaim, "How interesting, let me see it", but what you often really mean is "Let me feel it".

The hands have one further important function: they are part of our communication system; and in the extent to which they are used to communicate, not only words but emotions and ideas, the hands of man are unique in the animal world. Among the primates only chimpanzees appear to use their hands to communicate.

One hundred and forty-five years ago the Right Hon. Francis Henry, Earl of Bridgewater, left in his will the sum of £7,000 to the Royal Society. He requested that the money should be used for the purpose of sponsoring a number of treatises; one of these famous Bridgewater Treatises entitled *The Hand*, by Sir Charles Bell, was a very remarkable study in the adaptation of the hand of man and animals, the more so since Bell was writing at a time when, probably due to the persisting influence of the French naturalist, Buffon, the "imperfections" of animals were a popular device with which to demonstrate the "perfections" of man; the sloth, for instance, was much pitied for "its bungled and faulty composition". Those familiar with the absurd anatomy of the sloth will appreciate the mistake, but actually the animal is ideally built for its role in nature. Bell, however, did not fall into this trap; he was clearly aware of John Hunter's principle that "structure was the intimate expression of function" and that *function* was conditioned by the environment. Lacking the instruction of Darwin (who was at that very time about his business on HMS *Beagle*), Bell was unaware that the catalyst effecting the perfect correlation between structure and function was not a matter of special creation but a matter of natural selection. His thesis was subtitled "its mechanism and vital endowments as evincing design". To Bell, the "design" owed more to divine inspiration at a celestial drawing-board than to the trial and error of an earthly workshop. But Sir Charles Bell was a functionalist who appreciated, above all, the relationship between the form of animals and the world in which they lived. In his belief that the mechanism of this relationship was "special creation" he was merely reflecting the scientific convictions of his time.

It is difficult today to realise how great was the impact of Darwin's theory on the fundamentalist way of thought. Darwin's *On the Origin of Species* was received with amusement by some, with profound scientific doubts by others, but with thunderous and righteous wrath by the Church, notably in the person of the famous Bishop Wilberforce.

I suppose we all have our heroes. I have three: Hunter, Bell and Darwin. John Hunter turned our attention from the structure of the hand to its function; Bell related the function of the hand to the environment; and Darwin demonstrated that the environment, by process of natural selection, gave birth to structure. The manner in which an animal responds to its environment is called its behaviour; therefore in terms of animals, human and otherwise, we have three levels at which it is necessary to consider any biological problem: structure, function and behaviour. Structure, the physical basis; function, the integrated action of the physical parts; behaviour, the deployment of function in the setting of the environment. It is at these three levels that I shall be looking at the hand in this book.

The success of an organism ultimately depends on its behaviour. In automobile language there is no virtue in a car that doesn't corner or whose petrol consumption is impossibly extravagant. Competition from other cars would soon drive it into obsolescence. In nature the same thing applies to animals. This is in fact the basis of the theory of natural selection. Pursuing the car analogy, natural selection is replaced by consumer choice; no one is going to buy a car which looks and sounds good when ticking over (function) but is incapable for one reason or another of behaving properly in its natural environment – the road.

Man's hand shows an extraordinary degree of primitiveness, an astounding conclusion when one thinks of its specialised movements, its acute sensitivity, its precision, subtlety and expressiveness. Can the hands of Leonardo da Vinci, a Swiss watchmaker or a Balinese temple dancer really be called primitive? There is an explanation of this apparent paradox between specialised and primitive. The hand itself is derived from yeoman stock but the factor that places it among the nobles is, as it were, its connections – its connections with the higher centres of the brain. Should the nerve supply be severed by injury or disease, the hand is left high and dry, isolated from its specialised organ of control. The disenfranchised hand is like a beautiful Rolls-Royce that has no engine under the bonnet – elegant but useless. Thus it is possible to argue that while the nervous system is very specialised in man, the hand, which retains many ancestral features and has acquired very few new ones, is extremely primitive.

I would like to end this brief introductory chapter with two stories, one of which is true and the other is probably apocryphal. Both, however, make the valid point that in an emergency the hand is the best surgical instrument ever invented.

The true story concerns myself. As a newly qualified doctor I was confronted one evening with a young girl with appalling injuries to her right leg as a result of a motor-cycle accident. When she was brought into the reception bay of the hospital, I could see the main artery of the leg completely severed in the thigh, pumping out blood at a terrifying rate. There were no instruments immediately available and no blood transfusion. So I used my hand as a clamp, pinching off the artery with thumb and fore-finger as well as I could. Finally I got a bit of string, all that was available, round the artery and tied it off. The blood stopped pumping. It would be nice to report that this unorthodox procedure saved her life, but sadly she died within a few hours. The point of the story is that nothing but the hands could have dealt with that emergency so quickly and effectively. Few patients, thank heaven, ever realise how, during an operation, an appropriately placed finger has saved their lives.

The apocryphal story is about a bluff and arrogant surgeon who was expanding on his favourite topic. "The hand", he said, with his usual pomposity, "is the surgeon's finest instrument." A physician overhearing him, remarked *sotto voce*, "At least the old bugger won't be able to leave *that* behind in the abdomen!"

Chapter Two
Structure of the Hand

Structure of the Hand

I have tried to avoid too many technical terms in this book as they make uncomfortable reading but there are a few which are almost obligatory to any discussion of the hand: the *phalanges* (singular: *phalanx*), *metacarpals* and *carpals*, *thenar* and *hypothenar*, *palm* and *dorsum*, *proximal* and *distal*, and *ulnar* and *radial*.

The *phalanges* are the bones of the fingers, *metacarpals* the bones of the palm, and *carpals* the bones of the wrist (Fig. 3).

Thenar describes the eminence and the muscles comprising it on the thumb side of the hand and *hypothenar* refers to the muscular eminence on the little finger side. *Palm* and *dorsum* (or palmar and dorsal) indicate the front and back of the hand respectively; they are the equivalents of *anterior* and *posterior*. *Proximal* and *distal* are general anatomical terms which denote the relative position of two structures or regions to each other. Proximal is the nearer of the two to the centre of the body in terms of the limb as a whole in the "anatomical position", while distal is further away. Thus the palm is proximal to the fingers and distal to the wrist. The fingers consist of three bones (phalanges) – the proximal, the middle and the distal. The *ulnar* and *radial* borders of the hand are named after the two bones of the forearm, the ulna and the radius. When the arm is in the so-called "anatomical position" in which the palm faces forwards, the ulnar border lies on the inner side and the radial border on the outer.

The idea of the "anatomical position" underlies all terminology in human anatomy, much to the confusion of zoologists whose subjects do not stand upright on two legs. The concept of an anatomical position is derived from the posture in which human cadavers were in Hunter's time suspended for dissection, either for reasons of space or verisimilitude. In this position the abdomen faces forwards and the back faces backwards. The palms of the hands are rotated so they face forward, but the soles of the feet face backwards. So, the palm of the hand is said to be directed "anteriorly", but its homologue the sole of the foot "posteriorly". Anatomical nomenclature is, to say the least, somewhat confusing. The inward twist of the limbs that takes place during development of the embryo makes the future plantar surface (the sole of the foot) face downwards. This occurs in both sets of limbs, but in the upper limb in many species the inward twist is capable of being corrected during life by means of a free movement between the two forearm bones; the hand can be held palm-upwards (*supination*) or back-upwards (*pronation*). The foot is held in

3. Bones of the hand seen in relation to the soft tissues.

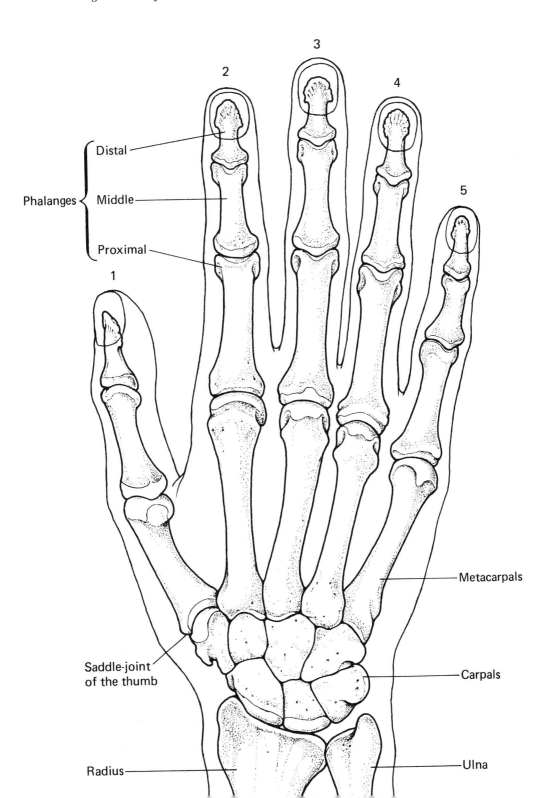

Phalanges { Distal

Middle

Proximal

Saddle-joint
of the thumb

Radius

Metacarpals

Carpals

Ulna

permanent pronation and cannot be corrected. The whole structure of the hindlimb in man and other mammals is suborned to the needs of weight-bearing.

The Digits

The Number of Digits

The late Professor F. Wood Jones called the human hand the "absolute bed-rock of mammalian primitiveness". He was referring particularly to the five-fingered state (*pentadactyly*). No mammal, indeed no reptile, amphibian or bird has ever evolved a form with *more* than five digits. There are many which have less, but none which have ever managed to scrape up an extra finger or two as a fixed characteristic of the species. Among the last shots in the war about polydactylism, which was much enjoyed by many mid-Victorian biologists, was a paper by two scientists who in 1892 showed that the multi-toed condition in the Dorking Fowl was in fact due to an abnormal cleavage of the developing limb bud and was not genetically induced. Strains of six-toed domestic cats, having the same genetic background, are common enough. A rather special case of polydactylism was suspected in the Giant Panda but Wood Jones (1942) demonstrated with X-rays that the panda's "thumb", which provides such a valuable device for clutching bamboo shoots (Fig. 4), was simply an enormously enlarged wrist bone called the radial sesamoid. The Giant Panda's thumb is capable of restricted and very simple movements.

Polydactyly is fairly uncommon in man as is *syndactyly* in which adjacent digits are united by webs of skin. The frequency of this congenital abnormality is between one and three per 1000 normal births, with a slight preponderance in favour of males (Fig. 5). Syndactyly, whilst pathological in man (Fig. 6), is normal for the Siamang gibbon, one of the ape family; in fact this characteristic is perpetuated in its scientific name, *Symphalangus syndactylus*. The second and third digits of the foot of the Siamang are permanently united by a web of skin.

Absence of digits (*ectrodactyly*) is another example of a situation which is abnormal in man but normal in certain primate species, and indicates how variability provides the raw material for natural selection to work on. What was initially an occasional congenital

4. *The false thumb of the Giant Panda used for clasping bamboo shoots while feeding (Courtesy of Field Museum of Natural History, Chicago).*

abnormality initiated during the development of the individual proves advantageous to those individuals possessing it and ultimately becomes a permanent fixture in the genetic code. Three examples of ectrodactyly are the commonplaces of primate anatomy. The spider monkey from the New World and the Colobus monkey from the Old are both thumbless, a condition that has nothing to do with zoological affinity of the two species but results from the evolutionary mechanism of parallelism, in which species with similar environmental needs tend to develop comparable physical characteristics. Another example of ectrodactyly is the loss of the index finger in the West African loris, the Angwantibo or Golden Potto. There is no special virtue in possessing an index finger if you are a potto, indeed there is a distinct disadvantage as it impairs the span of the grasp.

Ectrodactyly is one of the strange conventions of cartoonists for whom the hand consists of three or four digits only. This is true both of human characters and anthropomorphic animal figures. One can only suppose that this foible stems from the simple fact that the hand requires the extremely meticulous draughtsmanship of a Dürer or a Watteau if it is not to end up looking like a bunch of bananas. Cartoons essentially are quickly executed line drawings.

The Length of Digits

The term *digital formula* refers to the relative projection of the digits when the hand is laid flat upon the table. The expression was a neologism of Wood Jones' and he explains the occurrence of digits of different lengths in the following lucid paragraph in his book *Arboreal Man*.

> The developing limb of all [vertebrates] appears at first as a small outgrowth from the side of the body. This outgrowth, known as a limb bud, takes the form of a little fleshy paddle with a rounded free edge. It is from the curved free margin that the digits grow outwards in a radiating manner. It is obvious if all the digits are of equal length, the middle will project ahead of the others; those standing next on either side will not reach quite so far forwards, and the digits upon the extreme edge of the palm or sole will have their tips still further behind.

5. *Polydactyly in man (Courtesy of J. N. Barron).*

6. *Syndactyly in man (Courtesy of J. N. Barron).*

32

5

6

If the digits continued to develop as they began, the digital formula would be symmetrical (3 > 4 = 2 > 5 = 1)*, but the digits evolved differential functions in later life and the formula has departed somewhat from the basic pattern of the developing embryo: the thumb has become shorter than the little finger, and the index finger has become longer than the ring finger in a proportion of adults (see Fig. 3).

The third digit (or middle finger) is unequivocally the longest but there is a battle for dominance between the second (index) and the fourth (ring) digits. As it turns out it is a battle of the sexes. A study of 881 college students, predominantly male, revealed that the index was longer than the ring in 28·6 per cent. When female records were removed from the series the score was lowered to 22·2 per cent, which suggests that females play a large part in keeping the percentage of long index fingers high. When a series of 252 female students was investigated it was seen that the percentage figures shot up to 45 per cent, confirming the fact that a long index finger was a characteristic of females.

No such dimorphism exists in primates so far as I know. In my own case, I appear to be bi-sexual. I have the male formula on my right hand and the female arrangement on my left.

It would be most inadvisable to draw any profound conclusions from the sexual difference in index dominance. Wood Jones, without committing himself in any way, merely labels the phenomenon as another unique and progressive specialisation of man (zoologically speaking) that emphasises the independence, or "aloofness" as he calls it, of man's index finger. There is no evidence that relative index finger length has any effect on skill.

Numbers of Bones in the Hand

The retention of a primitive number of bones in the fingers and thumb is hallowed by time. The phalangeal formula of 2 . 3 . 3 . 3 . 3 ., reading from thumb (which has two) to little finger, is the pattern shared by all mammals, living or extinct, and even some mammal-like reptiles, the forerunners of the true mammals. The ancestral amphibian hand exemplified by the fossil *Eryops* displays a phalangeal formula of 2 . 3 . 4 . 5 . 3 ., a pattern that has persisted among reptiles and the majority of mammal-like reptiles. It is interesting to

*The symbols > and < signify "greater than" and "less than" respectively. The equals sign indicates that the digits are the same length.

note that the dissident digits are the third and fourth, the leading digits in terrestrial locomotion of all four-limbed creatures. In two-toed ungulates with cloven hoofs the persisting digits are also the third and fourth. It is also notable that the thumb has consisted of only *two* bones for a very long time, indeed it is questionable whether it has ever possessed more than two phalanges, a conclusion which will no doubt infuriate the legions of anatomists who collectively have expended much time and energy trying to determine whether the missing elements of the thumb is a phalanx or a metacarpal.

The *carpal bones* are eight in number in most living reptiles and in all mammals. The carpals in primates consist of two rows of bones and an intermediately positioned bone called the centrale. From the thumb side the bones of the distal row are called the trapezium, trapezoid, capitate, hamate; the bones of the proximal row consist of the scaphoid, centrale, lunate and triquetral. In certain orders, reductions and fusions have taken place; for instance the scaphoid and the lunate of human anatomy are fused in carnivores to form a single element, the scapho-lunate.

The primate carpal formula is as primitive as it could be, consisting of eight bones arranged in two rows. There are no departures from this theme except in the case of man who has no centrale bone; the carpus therefore is reduced to seven bones (Fig. 3). This distinction between man and the rest of the primates should delight those whose concern is to collect examples of absolute differences between man and monkeys in the cause of fundamentalism or some such quirk. Sadly for them the uniqueness of man in this respect is purely illusory because the centrale bone, strictly speaking, isn't missing at all; it is present as a discrete cartilaginous nodule in the very early stages of intra-uterine development. Subsequently it becomes fused with the adjacent scaphoid and is no longer discrete.

A similar fusion of the centrale occurs in the chimpanzee and gorilla; the only distinction between them and us being the date at which fusion takes place – well before birth in men and at varying times just before or just after birth in the African apes. Fusion has been accelerated in the case of man compared with the apes, the actual mechanism is however identical.

Wood Jones associates fusion of the centrale bone in man with the increasing independence of his thumb and index finger which requires a stable platform from which to operate; by fusion, one element of the chain is dispensed with. Fusion of the centrale bone in gorillas and chimpanzees has a different history. Both these apes are "knuckle-walkers", that is to say they bear the weight of the front part of the body on the backs of the knuckles when walking quadrupedally (Fig. 7). This special posture requires a number of bony adaptations to strengthen the metacarpals, carpals and wrist joint, including a reduction in the number of mobile bony elements in the carpus.

7. *The "knuckle-walking" posture used by chimpanzees and gorillas on the ground. Skin over the backs of the knuckles becomes hard and calloused.*

Modern	Alfred's Anglo-Saxon	Canute's Medieval Latin	Synonyms
Thumb–1st Digit	Duma	Pollex	–
Index–2nd Digit	Scythe finger	Demonstratorius	Forefinger, Salutorius, Shooting, Sagittator
Middle–3rd Digit	Midlestafinger	Impudicus	Medius, Famosus, Obscenus
Ring–4th Digit	Goldfinger	Annularis	Medicus
Little–5th Digit	Lytlafinger	Auricularis	Minimus, Pinkie

8. *The names of the digits.*

Terminology of the Digits

The naming of the digits, according to Bertelsen and Capener, can be traced back at least as far as Aethelbert, the first Christian Anglo-Danish King of Kent, who in AD 616 laid down a set of laws of compensation for the loss of fingers or thumb. King Alfred and King Canute, both thoughtful – if preoccupied (Alfred) or optimistic (Canute) – sovereigns, revised these laws and in so doing identified each digit by name.

King Canute's rates of compensation for the loss of digits and the scales laid down by the Department of Health and Social Security are a surprisingly close match. They range from 30 per cent for the loss of thumb (DHSS) and 30 solidi or shillings (Canute), to 7 per cent for the loss of the little finger. Clarkson and Pelly remark that compensation was surprisingly high (18 solidi) for the ring finger compared with 7 per cent on the DHSS scale. They suggest the ring finger may have had a religio-mystical significance. There is good reason to believe that the solidus represented a percentage.

The terms used by anatomists in digital identification are fairly obvious (Fig. 8). *Auricularis* (little finger) denotes the digit most commonly employed to extract wax from the depths of the outer ear.

The implication of *demonstratorius* (index finger) is self-explanatory, but why *impudicus* for the middle finger? Possibly because it was used as a gesture of derision in Latin and some Arab countries, just as the two-fingered jerking gesture carries a clear insult elsewhere; perhaps because being the longest digit it is ideally disposed to carry out sexual caresses of the female genitals or indelicate scratching operations.

Professor Wood Jones, a great authority on the hand, preferred the term *obscenus* (middle finger) to describe the digit that is used to express scorn and derision. The ring finger, the *annularis*, is again self-explanatory though its synonym, the once widely used *medicus*, is not; perhaps it is related to the fact that persons formally admitted as doctors of medicine used to wear a gold ring on that finger. One suggestion that has been made is that this digit was used by mediaeval physicians to stir their cordials and nostrums. Perhaps a latter-day survival of this custom is the trick that generations of consultant physicians have played on their students. The consultant, poker-faced, enquires of the student whether he has tasted the patient's urine. The student naturally admits that he hasn't, whereupon the consultant dips his finger into a specimen jar and touches it to his tongue. What the student, who somewhat unhappily follows his example, is unaware of is that it was the ring finger that the consultant dipped into the urine but the middle finger he put to his mouth.

The intruder (Fig. 8) is the "pinkie", a synonym for the little finger, a term widely used in the United States. Its derivation is obscure, but it possibly derives from a "pink" or "little boat", a word of Scottish origin. The modern terms are the thumb, the index, the middle, the ring and the little digits; however there is no reason why "finger" shouldn't be used instead (e.g. index finger). "Ring digit" is distinctly clumsy. Numerical identification using terms such as "second finger" or "first digit" (the thumb) should be avoided as being confusing.

Shape and Proportions of the Hand

It only requires a little observation during periods of ennui while travelling in trains and on buses to conclude that the human hand comes in all sorts of shapes and sizes. The English language is rich in adjectives commonly applied to the appearance of the hand: from beautiful and elegant to powerful and cruel. To describe the texture of the skin we can choose between coarse, velvety, smooth, hairy, horny, and moist. To describe its colour we have the inevitable lily-white at one extreme and brick-red at the other. To describe function we tend to resort to "professional" analogies: a surgeon's hand, a musician's hand, an artist's hand, a navvy's hand. We also talk of handshakes as being vice-like, flabby, cool, friendly, or warm. Hands are things that people notice in others and on which they place great reliance as indicators of character. They knew (so they say) that so-and-so was not to be trusted immediately they shook hands with him. Like all generalisations, this sort of typological thinking can lead one into serious errors of judgement. The most brilliant plastic surgeon I know has a short, powerful "square" type of hand that would have been no mean model for the power-orientated fists of our caveman ancestors.

The hands, like the face, are prime areas for the expression of individuality. Individuality is a non-technical concept which by its very heterogeneity defies classification. The equivalent scientific concept is variability. Variability is the raw material of natural selection, it is what nature selects from. It can express itself in structure and function but, probably more importantly, in behaviour. There is an old Yorkshire saying that there is "nowt so queer as folk" which means that one should not be surprised at anything humans do.

Variation in the shape of the hand is particularly obvious. Although function is not unduly dependent on shape, there are undoubtedly some professions which benefit from certain characteristics. The obstetrician is at an advantage in his profession if his hands are long, slender and flexible (the *main d'accoucheur*) as indeed is the concert pianist, and the magician whose ability to back-palm half a pack of playing cards depends upon it; he also requires a long little finger, which is a variable feature in man. Alternatively the broad, stubby, muscular hand is much more suited to pick and shovel work for which an extensive gripping surface is an advantage.

For all its variation in shape, the overall function of the hand is universal. Some individuals may be "good with their hands" and others "all fingers and thumbs", but the basis for skill doesn't lie in the hand. There are rogue cars that have to have engines replaced within the first few weeks of ownership, but there are no rogue hands. The skill of the hand lies in the brain and it is here that dexterity and

Structure of the Hand

Sample	Number	Sex	Mean of hand length (mm)	Mean of hand breadth (mm)	Shape Index HAND BREADTH × 100 / HAND LENGTH
German Civilians	1200	M	188·0	86·0	45·7
US Army Drivers	2683	M	193·0	88·1	45·6
USAF Flying and Training Personnel	3000	M	190·0	88·0	46·3
Flying Trainees	3000	M	192·0	87·0	45·3
US Army Recruits	8000	F	175·0	77·0	43·9
WAF Trainees	848	F	172·0	77·0	44·8
Total	18,731 Males and Females				Range 43·9–46·3

Sources of Samples: Brezina & Lebzelter, 1953; Daniels & Meyers, 1953; Daniels, Meyers & Worrall, 1953; Herzberg & Daniels, 1953; McFarland & Damon, 1953; Randall & Munro, 1949.

9. Shape index of human hands.

adroitness (and clumsiness) originate. The hand is a mirror of the brain, therefore there can be no such combination as dextrous hands and clumsy brains.

I have analysed the data from many published sources based on measurements of 18,731 individuals (8,480 females and 10,251 males) and it is apparent that the overall *shape index* (the ratio of length to breadth) is remarkably constant (Fig. 9). Furthermore the proportions of the male and female hands differ only very slightly (absolute dimensions apart), the female being on average only a shade more slender than the average male – a rather surprising result. All the data on women stem from the US Army Services and it is possible therefore that there may have been some bias in favour of the robust type of girl with what are known as "capable" hands. Perhaps the Dresden shepherdess-types were weeded out at their medicals. Had data been included from India, Sri Lanka, Burma and points East where women's hands are notoriously slender, the results might have been very different.

Among a number of indices that can be derived from available data on the skeleton of the hand, one, the *phalangeal index*, is particularly valuable in palaeontology where there are only bones to go on. It provides insights into the locomotor behaviour of fossil apes and monkeys and suggests that the proportions of the human hand have been on the evolutionary production line for at least twenty million years.

The phalangeal index expresses the length of the three phalanges of the middle finger as a ratio of the total hand length. Functionally it reflects in primates the degree to which the hand is adapted for grasping and climbing in an arboreal milieu, or walking and running in a terrestrial one – the rationale being that the longer the fingers the greater is the capacity for grasping; and the shorter the fingers the more stable and powerful is the lever in quadrupedal ground-walkers.

In man the phalangeal index has a mean value of 49 per cent. This figure is intermediate between that of ground-adapted monkeys such as the baboon, the patas monkey and the gelada (42–44 per cent), and the arboreal specialists (52–58 per cent). Man shares the 48–50 per cent bracket with part-time arboreal Old World monkeys like the macaques, mandrills and the mangabeys which are intermediate in their ecological preferences. With a few exceptions the lemurs and lorises (collectively known as prosimians) have the longest digits of all, the Index range showing wide variation with values of 55–64 per cent. The long fingers and short metacarpals of arboreal monkeys and prosimians are primitive retentions from the ancestral stock in which long fingers was characteristic.

The hands of the three great apes are specialised in many ways but particularly in respect of their arm-swinging type of locomotion known as *brachiation*. Although they possess remarkably long fingers, the metacarpal and carpal sections are equally elongated, having kept pace with the fingers during evolutionary growth. Consequently the Index value approaches 50 per cent and is thus very similar to man's: gorilla 48 per cent, chimpanzee 50 per cent, orang-utan 50 per cent.

In practice it is true to say that the living gorilla is almost wholly ground-living. The male never climbs trees and the juveniles and females do so only to a very limited extent. However the grounded gorilla does not adopt the typical quadrupedal gait of monkeys but instead has opted for a *knuckle-walking* posture (see Fig. 7) whereby contact with the ground is made by the friction pads on the back of the middle phalanges of digits 2–5. Consequently there have been no evolutionary pressures tending to shorten the fingers because there would be no particular advantage in so doing. The same is true of the chimpanzees which are slightly more aboreal in habit. Man never became a knuckle-walker as did the gorillas and chimpanzees, but his ape-like ancestors, eschewing a monkey-like quadrupedal gait in favour of bipedalism, also avoided excessive shortening of the hand, the fingers in particular.

An interesting mechanism called "double-locking" occurs in chimpanzees and orang-utans. Owing to the exceptionally long palm, the tips of the fingers can be tucked into the skin-fold at the base where the finger meets the palm. This action is difficult to describe but is clear enough in the figure. With further flexion the locked fingers are rolled into the palm (Fig. 10), thus "double-locking"

10. Double-locking. Chimpanzees and orang-utans possess long fingers but proportionately longer palms; as a result they are able to tuck the tips of the fingers into the crease at their bases as shown in this diagram. Further flexion towards the palm constitutes double-locking (see Fig. 11).

11. Mechanism called double-locking facilitates a powerful grip of slender food-bearing branches or lianes.

them. This mechanism is somewhat analogous to an engineer's over-centre lock. The adaptive value of double-locking is apparent during feeding. While feeding, both chimpanzees and orangs prefer utilising the slender branches of the trees. Both animals use their feet as well as the hands to distribute the weight evenly between several slim, flexible supports. It is obvious that for slender branches the most effective type of grip possible is in the double-locked position.

The power of the orang's grip by means of double-locking can be demonstrated with a piece of string. A man cannot resist a direct pull along the length of the string, but orangs and chimpanzees have no such problem (Fig. 11). The moral here is never take on a tug-of-war team that has chimpanzees or orang-utans on its side.

Skin

The skin surface of the human hand, far from being smooth and perfect like alabaster, as so often in romantic novels, is neither smooth nor white but highly textured and multi-coloured. It is deeply etched with lines, grooves, folds and furrows and studded with pits for hairs and open ducts for the excretion of sweat. In living hands veins show blue, arteries show pink and for good measure freckles and "liver spots" show brown; only dead hands are lily-white.

The role of the skin in general terms has been epitomised in somewhat bluff, non-technical language as the structure that "keeps the blood in, and the rain out". Never mind, the axiom, though crude, provides a starting point. Besides being water- and blood-proof, the skin

> . . . becomes thick where it is subjected to rough treatment, it is fastened down where it is most liable to be pulled off, it has friction ridges where it is most liable to slip. Even with our ingenious modern machinery, we cannot create a tough but highly elastic fabric that will withstand heat and cold, wet and drought, acid and alkali, microbic invasion and the wear and tear of three score years and ten yet effect its own repairs throughout and even present a reasonable protection of pigment against the sun's rays. It is indeed a fine fighting tissue. (S. Whitnall quoted by J. C. B. Grant in *A Method of Anatomy*, 5th Edition, p. 70.)

The skin of the palm is firmly bound down to the underlying, packing, tissue of the hand. There are areas, over the ball of the thumb and the "heel" of the hand, where the skin is relatively mobile. This is because the underlying fibrous tissue, which is heavily loaded with fat, forms a distinct pad. But elsewhere – as in the central area of the palm – the underlying tissues are fat-free and the skin is firmly adherent.

There was a period during the early months of the last war when among naval gunnery personnel, severe flash burns of the palm of the hand were rather common. They were treated by a full-thickness skin graft (a full-thickness graft includes the fatty subcutaneous layer as well as the skin). The early post-operative results were encouraging but when the patients started to use their hands again for grasping, the lack of the normal fixation of fat to fibrous tissue in the centre of the palm meant that the grip was slippery and ineffective, and objects thus gripped were flying about like orange pips squeezed between finger and thumb. The lesson was quickly learned and the fatty layer was not included in subsequent skin-grafts for this purpose.

Three types of lines can be distinguished in the hand: *flexure lines*, *tension lines* (Langer's lines) and *papillary ridges* (Fig. 12).

44

FLEXURE LINES are in effect skin hinges which open and close during grasping and gripping, following the movements of the underlying joints. The flexure lines are permanent creases and are analogous with the concertina-like folds that develop in a well-worn jacket across the elbow region. They are lines of skin stasis produced by an anchoring of the skin to the deep packing tissues. Because of their relationship to underlying joints, flexure lines define the principal axes of movements of the hand. However, there is a great deal of

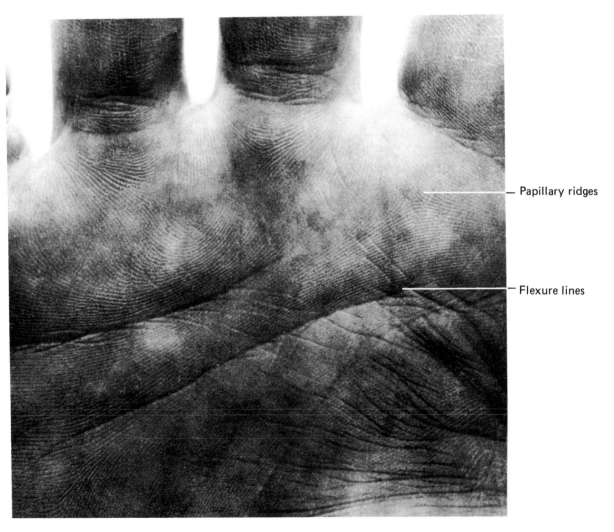

— Papillary ridges

— Flexure lines

12. Palm of hand showing flexure lines and papillary ridges.

variability. There are also a number of subsidiary flexure creases which vary in position in individual hands; subsidiary creases "take up" the excess skin that balloons up when the digits are folded.

The infinite variety of primary and subsidiary flexure lines is a source of great satisfaction to professional palmists whose living is thereby guaranteed. Although there is no scientific support for the idea that palmar creases have any psychic or predictive meaning, it is well-known that hands show evidence of occupation, disease and to some extent of temperament and personality – quite an adequate battery of information for the astute palmist to build up a pretty impressive augury. Professor J. D. Lever wrote in 1963:

> In spite of the sceptics [in Roman times], palmistry was considered a worthy subject for study by the intellectuals, and there is the story . . . that Caesar, who prided himself on his infallibility as a palm reader, denounced as an imposter one who appeared before him as Alexander, son of Herod, because his hand did not bear the royal markings. Biblical support for chiromancy can of course be found in the Book of Job (37, 7) . . . He sealeth up the hand of every man that all men he has made may know it.

It is not surprising that with such high level endorsement the ancient art has openly flourished in every age. In our own, palmistry has a somewhat eccentric, though faithful, following.

The most conspicuous horizontal creases in the palm are the *heart line* which extends from the inner or ulnar border of the hand to the region of the cleft between the index and middle finger, and the *head line* which starts in the middle of the palm and curves toward the radial border of the hand paralleling the heart line for a short distance but eventually passing beyond it (Fig. 13). In non-human primates the head and heart lines are not separate, rather they form a single crease which runs straight from ulnar to radial border of the hand without a break. The difference between the simian type of horizontal crease and the human heart-head lines reflects the fact that in man the index finger has a great deal of independent mobility, whereas, in non-human primates, the index finger tends to work in harness with the remainder. A "simian crease" is a fairly rare abnormality in man, cropping up in about 0·4 per cent of the population (Fig. 14). Wood Jones summed up the difference between simians and non-simians in the following terms: "The transverse creases of the human palm are

46

13. Hand print showing main flexure lines of palm, named in palmist's terms.

14. "Simian crease" apparent on the right; normal on left.
Note 3. 4. 2. 5. formula on both sides (Courtesy of St Thomas's
Hospital Medical School).

distinguished from the corresponding creases in the hands of
monkeys and anthropoids by their curved form and oblique
direction." (Fig. 27).

Among the other flexure lines that constantly appear are those
related to the finger joints which may be double or even triple; it is the
distal crease, or – in the case of a triple arrangement – the middle line
which most closely marks the line of the joint. The proximal digital
crease is usually single in the little and index fingers and double in the
remainder; this difference reflects the extra mobility of the two
marginal fingers.

The major palmar creases consist of two consistent horizontal lines
(described above) and two or three vertical lines (Fig. 13). The

vertical line nearest to the thumb is called the *line of life* by palmists and is the first to develop in the embryo. The second vertical line which splits the palm into ulnar and radial segments is the so-called *line of destiny*. These two lines are the principal skin hinge for the thumb.

There are usually two bracelet creases at the wrist. The distal crease, however, which marks the end of the palm, may break up into a double track giving the appearance of three skin "bracelets". The proximal bracelet corresponds to the wrist joint and the distal bracelet corresponds closely to the transverse axis of the joint between the two rows of carpal bones (mid-carpal joint).

Flexion creases and papillary ridges are especially prone to show abnormalities resulting from growth disturbances in the developing foetus, particularly in cases of genetic abnormality. For instance in Mongolism (now known as Down's syndrome) a "simian crease" is often present and the full complement of flexion creases are absent from the little finger.

TENSION LINES provide the basis for the wrinkling of the skin which is such a source of anxiety to the middle-aged who still regard themselves as youthful. Tension lines occur all over the body but it isn't until the skin loses its elasticity with advancing years that they form permanent wrinkles. The function of these lines is to provide the skin with a certain amount of "stretchability" in directions corresponding with the natural demands of the region. On the back of the elbow for instance, the tension lines are arranged horizontally, while on the back and on the abdomen they are omnidirectional. Bundles of fibrous tissue interspersed with elastic fibres provide the extensile element over most of the body, but in certain areas, such as the face and the scrotum, muscle fibres are attached directly to the skin. Pronounced wrinkling is usually due to muscle activity. Some wrinkles are congenital while others, particularly on the face, are acquired (or at least exaggerated) by a lifetime of muscular activity associated with certain facial expressions. "Crows' feet" are a perfect example of tension lines which form with age.

Tension lines in the hand are seen as fine, concertina-like folds arranged horizontally on the dorsum, obliquely on the thenar eminence and vertically on the phalanges. Try pinching up the skin on the back of your hand; either it snaps back into place with the celerity of an elastic band, in which case you are young, or it flattens out sluggishly, in which case middle-age is your problem, or it stays pinched up and you are what is known euphemistically as a senior citizen.

PAPILLARY RIDGES are permanent thickenings of the epidermis, the outer cellular layer of the skin. They are raised above the general level of the skin, and are only to be found on the palmar surface of the hand and the sole of the foot. Their distribution corresponds with the principle areas of gripping and weight-bearing where they serve very much the same function as the treads on an automobile tyre. To carry the analogy further, a slight dampness improves the grip of both hand and tyre, while wetness leads to a loss of grip in both. Although this is the most popular view, it isn't wholly true that the function of papillary ridges is purely mechanical; their role as outriders for the sense of touch is equally important, perhaps more so.

Ridges vary in coarseness on different parts of the hand. They are at their finest on the finger tips and at their coarsest on the rest of the fingers, the palm of the hand being intermediate. Individually there is a certain amount of variation; there is, for example, a loose correlation between ridge width and body size. A giant has ridges of giant proportions and a midget ridges of midget proportions but the correlation for intermediate statures is not so strong. Professor Harold Cummins has suggested that there are two genetic factors controlling ridge size: the factor controlling body size generally and an independent factor for ridge width.

Papillary ridges are rather like an iceberg: there is as much under the skin as protruding above it. The inner portion consists of a number of finger-like processes intimately related to the organised nerve endings of the epidermis, particularly those that serve touch, which are called Meissner's corpuscles. The papillary ridges act as little lever arms and transmit the most delicate contacts and pressures to the underlying finger-like processes which are displaced in their turn and so stimulate the Meissner's corpuscles and other endings such as Merkel's disc and free nerve endings.

On the surface the papillary ridges are arranged in parallel formations, sometimes in curved series, sometimes straight. The visual analogy between papillary ridges and a contour-ploughed field is very close, but the texture of a piece of corduroy with its uni-directional friction effect provides a better mechanical analogy. However, neither of these analogues includes the third and most important feature which complements both the mechanical and sensory functions of the papillary ridges, and that is sweating.

Sweat glands of the common or garden variety, known as *eccrine* glands, are found all over the body in man, but in non-primate mammals they may be absent from hairy skin and are only present in the hairless walking pads of the extremities. Non-primate animals have *apocrine* glands, which are very similar to sweat glands in structure but not in function. Instead of the watery consistency of sweat, apocrine glands produce a thick and viscid fluid that is odourless at first but rapidly becomes rancid as the secretion is decomposed by bacterial action. The typical smell of animals is, at least in part, a function of their apocrine glands. Although body odours are suppressed in modern western society by means of perfumes and aerosol deodorants, man has not always been so prissy. Time was when the village swain would place an apple in his armpit to offer to his partner for immediate consumption at the end of a vigorous barn dance; he clearly was intuitively aware of the strange and potent effect of cheap music and sweaty apples, as Noël Coward might have said.

Among the primates there is a trend for the eccrine glands to increase at the expense of apocrine glands as we ascend the evolutionary scale towards man. Prosimians have no eccrine glands at all except in the palms of the hands and the soles of the feet; monkeys have a few, chimpanzees and gorillas have more eccrines than apocrines and, finally, man has an enormous preponderance of eccrine glands amounting to 150–340 per square centimetre of skin. Man is a very sweaty animal.

Eccrine glands lie just below the epidermis, forming coiled cellular tubes; a single duct opens to the surface where they can be easily seen with a hand lens, particularly when they are secreting. In the hand the sweat-duct openings are aligned along the crest of the papillary ridges (Fig. 15). The papillary ridge therefore has a built-in lubricating system, which keeps the skin moist and adhesive and materially enhances the grip of the hand. The dry hand provides a poor gripping surface, as every workman knows who has ever spat on his hands before grasping a pick-handle.

Lubrication of the hands also improves the sense of touch. Both pain and touch are more exquisite in the presence of damp skin. The effect of sweat is that the papillary ridges swell and become more elevated, thus increasing their effectiveness as contact points. By far the greatest concentration of sweat glands occurs in the palm of the hand and the sole of the foot. These glands are particularly interesting. Although sweat glands of the hands are structurally identical with glands elsewhere in the body, they have a slightly different development history; furthermore, they cease activity during sleep and they respond not to *thermal* but to psychic stimuli under stress, a function believed to be mediated by hormones. All three recognised functions of the papillary ridges have at one time been regarded as the most important, but it is a fruitless argument and not worth pursuing. Speaking teleologically of course, you might say the principal function of papillary ridges was to catch criminals.

Hair, Nails and Fat Pads

Desmond Morris has a lot to answer for. No one was too worried about hair and its phylogenetic significance in apes and men until he put the cat among the pigeons by calling his provocative and important book *The Naked Ape*. It immediately raised a host of questions, one of which was: is man really hairless? There are two ways of answering this, anatomically and physiologically. In spite of appearances, the number of hairs per unit area in chimpanzees and humans is much about the same. The apparent difference lies in the

15. Tip of the finger showing papillary ridges. Note sweat pores along crest of each ridge.

16. (Facing) Male and female Europeans have approximately the same number of body hairs. Misleading differences between them is that the male's body hair is long, dark and coarse; female's is short, fine and colourless (Courtesy of Drs W. Montagna and R. J. Harrison).

quality of the hair. Human body hair is fine, short and pale while the chimpanzee's is coarse, long and dark with melanin pigment. There are slight differences in distribution, for instance a Caucasian has coarse facial hair which is lacking in chimps, and chimps have a lower hair-line on the forehead, a condition which is frequently met with temporarily in human newborns. Both men and chimpanzees are prone to baldness at the front of the head. Although male Caucasians appear to be more hirsute than females (Fig. 16), the number of hairs on the body is similar; once again the difference lies in the quality, not the quantity, of the hairs.

Thus humans are not naked in an anatomical sense, but from the point of view of physiology they are. Since the fine hairs afford no protection against heat loss or the harmful effects of the rays of the sun, they might just as well not be there. While fine pale hairs are useless relics from a hirsute past, the visible hairs of the forearms and legs, for example, function in much the same way as papillary ridges in acting as contact points for the sensory nerves that form a network around the hair follicle.

Although it wouldn't have had such a provocative ring, Desmond Morris might have been more correct to title his book *The Fatty Ape*. One of mankind's great unsung hallmarks is the layer of fat interlaced with fibrous tissue (*panniculus adiposus*) that lies between the skin and the muscles binding them together and provides for both insulation and fat storage. Not even the apes are so blessed. In humans the distribution of subcutaneous fat is also a secondary sex characteristic (the female breasts for example). The *panniculus adiposus* is something that man shares, surprising as it may seem, with the whales, dolphins and porpoises whose subcutaneous fat gives them their sleek, smoothly contoured outline.

The hand is not particularly well endowed with hair; the palm of course is totally hairless and the hair on the dorsum is not very evenly distributed. It is better developed on the inner or ulnar side than on the radial and it is present in tufts on the proximal and middle phalanges but it is absent over the knuckles; it is usually absent on the back of the thumb. The hairy coat of mammals (and man) is not distributed haphazardly, it is arranged in definite tracts according to functional needs. All the hand hair sweeps towards the ulnar border. As Wood Jones points out, if you remember this you need never be at a loss again when your friendly neighbourhood policeman requests your assistance in deciding which side an isolated human finger comes from. It comes from the side to which the hairs are pointing.

The central part of the palm of the hand is relatively fat-free, which facilitates gripping; the thenar and hypothenar eminences, however, carry fat pads below the skin.

Right fore-paw

Rat

Right hind-paw

17. Two-handed feeding – indicating lack of prehensility – in some mammals (Courtesy of The Illustrated London News*).*

Nails

With the exception of marmosets and tamarins, small-sized South American monkeys, the tips of the digits in primates are surmounted by nails. Most mammals which do not have hooves have claws, sometimes sharp and scimitar-like as in cats, sometimes stout and blunted as in dogs. It is generally thought that nails have evolved out of claws opening out to form a relatively flat curved carapace to protect the fingertip. The actual shape of the nail is dictated by the bulk and curvature of the underlying fingertip. In some South American monkeys the tips of the digits are so narrow and the pads so poorly developed that the nails are steeply curved from side-to-side almost as if they had reverted once more to the more primitive claw.

Claws are not compatible with prehensility because of the mechanical obstruction of claws overgrowing the fingertips; occasionally a marmoset will grip its food between the tips of the claws and the proximal part of the palm but the action lacks any element of precision. Many clawed mammals, especially the smaller ones like squirrels and marmots, hold their food between their two paws as one would a beach-ball; in this way they get round their lack of single-handed prehensility (Fig. 17).

Human nails are composed of keratin in the form of flat compacted cells derived from the skin. They develop in the human embryo

Squirrel
grasping

Fore

Squirrel

Raccoon

Raccoon
left-hind

Raccoon fore-paws

between the eighth and twelfth week of foetal life. Nails themselves are dead tissue but are set on a plate of living flesh rich in blood vessels and nerve endings. The source of new growth is a matrix of germinal cells buried beneath the fold of skin at the base of the nail. As they push forward the cells from the matrix become keratinised.

The growth of nails is continuous. The speed of growth depends on a number of factors such as age and the state of health. Nails are said to grow faster on the side of the dominant hand; nail-biting, too, is considered to be a stimulating factor. The ambient temperature is also believed to have an effect on growth rates, being faster in summer than in winter. The thumb is the fastest growing nail and the little finger the slowest. To some extent the size of the half-moon at the base of the nail is a guide to the speed of growth – the larger the faster. All things considered, the overall growth of human nails is of the order of 0·1 mm per day.

The responsiveness of nails to a variety of conditions, in the internal as well as the external environment, have interested physicians over the years, and nails constitute an important, if unobtrusive, element of any doctor's examination. Nails are the barometers of ill-health. Longitudinal ridges, brittle nails, transverse grooving, the size of the half-moons, the colour of the nail-bed seen through the translucent nail, and white marks in the keratin, are all indicators of ill-health in one way or another. Any severe illness (physical or psychiatric) may leave transverse linear depressions on each nail. These are called Beau's lines.

The principle function of nails is to provide both a rigid backing and a protective carapace for the pulpy fingertip which is of such critical importance for the manipulation of small objects and the discrimination of textures. The secondary functions are legion – from winding a watch to opening a cellophane-wrapped package. Habitual nail-biting may have many sinister temperamental overtones, but a far worse deprivation is the loss to the nail-biter of a built-in tool-kit of cutters, pliers, scrapers and screwdrivers.

Palmar Pads

The primitive mammalian "hand", which is of course functionally used as a foot, bears a maximum of eleven pads. There are five apical pads one at each finger tip, four interdigital pads in the distal area of the palm, and a thenar and a hypothenar pad in the proximal palm. The hypothenar pad is often divided into two parts (Fig. 18). Each

18. Palmar surface of the hand of an Old World Monkey. Note palmar pads surmounted by papillary systems.

palmar pad is surmounted by skin which bears papillary ridges. This is true even of man whose hand still bears the traces of the primitive quadrupedal walking pads. Man retains the five apical pads on the ends of the fingers, three interdigital pads and a hypothenar pad. The thenar pad has disappeared, the pad or "mount" of the thumb being largely a muscular eminence. The interdigital pads are barely discernible. The remnants can be seen as small puffy swellings at the base of the clefts between the fingers. The hypothenar pad serves to cushion the pressure exerted by the handles of tools and weapons held in a power grip (see Chapter 4). Below the skin of the hypothenar pad and partly attached to it is a small muscle called the *palmaris brevis* whose function is very mysterious. It helps to "cup" the hand forming the "Cup of Diogenes" (named after the Greek philosopher noted for his simple mode of life) and serves to stiffen the hypothenar fat pad, thus giving it more rigidity for power gripping purposes. As might be expected, there is no comparable muscle in the hands of monkeys or apes; it would appear to be a *bona-fide* hallmark of mankind.

The hypothenar pad is homologous with the heel pad of the human foot. In fact this part of the human hand is often referred to as the "heel". Like all palmar pads, the heel pad is made up of lobes of fat encapsulated in a network of fibrous tissue. This type of adipose tissue is not merely a depository for fat like the abdominal wall, but a structure with an essential functional role, that of defending underlying tissues from injury.

When the starving internees of German concentration camps were finally freed in 1945, most of them had used up all their available stored fat, even from the walls of the heart. Yet in spite of the fact that the prisoners were little more than walking skeletons, their heel pads were intact, rounded and as fatty as ever, and (although I don't recall any precise mention of this) I suspect that their hypothenar fat pads also survived.

Internal Anatomy

With the glove of skin stripped from the hand and the internal structure revealed, one begins to understand what is meant by *multum in parvo*. In an incredibly small space, tendons, muscles, nerves, veins and arteries are packed as close as sardines in a tin. It is possible to wax quite lyrical over the functional effectiveness of the anatomical arrangements of the hand as Sir Charles Bell did: ". . . and we must confess that it is in the human hand that we have the consummation of all perfection as an instrument."

Joints

The skeleton of the hand is comprised of twenty-seven elements of which fourteen are phalanges or finger bones, five are metacarpals or palm bones and eight are carpals or wrist bones (see Fig. 3). All of them articulate with at least one other bone by means of a synovial joint. Briefly, the synovial joint (whose essential feature is its mobility) is the commonest joint in the body. Movements are controlled by muscles and ligaments; if there are four movements possible at any given joint then there are at least four sets of muscles, one for each. Without muscles, the limbs would be flail – as uncontrollable as a marionette without strings. Synovial joints are enclosed in a capsule lined by a moist, glistening synovial membrane

19. The figure depicts a hypothetical synovial type of joint found in man and all mammals.

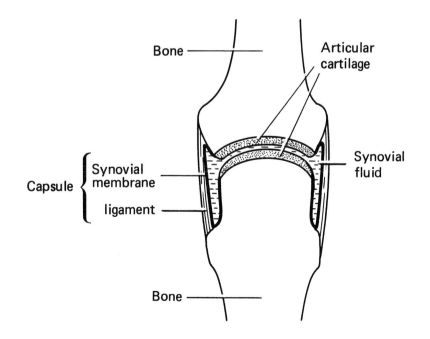

made up of cells which secrete the lubricating fluid which makes for the smooth action of a joint as well as supplying nourishment. The outside of the capsule is strengthened by ligaments which also act to some extent to guide the movements of the joint (Fig. 19).

There are several sorts of synovial joints which are classified according to the type of movements permitted. Joints with a single axis are of two kinds – the hinge and the pivot. The hinge-type operates like the lid of a box which opens and shuts moving about a single axis (the elbow is a good example); in the pivot joint, movement is limited to one of rotation around a fixed point (the skull and atlas vertebra pivot on the axis vertebra to allow twisting movements of the head on the neck). Bi-axial joints, or condyloid joints, are those that have two axes of movement at right angles to each other; poly-axial joints have multiple axes and they operate on the principle of a ball-and-socket. There is also rather a special joint – of immense significance to the working of the hand – called a saddle joint, which combines the properties of a bi-axial joint and a ball-and-socket joint. Finally there is the so-called plane joint, which allows only a sliding movement between the two opposed surfaces. In spite of its name, the surfaces of a plane joint are often slightly curved; the action of a plane joint can be mimicked by the movement of two stacked saucers sliding on one another.

All synovial joint types, with the exception of the pivot and the ball-and-socket, are represented in the hand: hinge-joint – phalanges articulating with each other; bi-axial joint – phalanges articulating with metacarpals; saddle-joint – the thumb metacarpal articulating with the carpal bone (the trapezium); plane-joint – joints between adjacent carpal bones.

Movements of the Fingers

The fingers can be bent (flexed) or straightened (extended) at the hinge joints between the phalanges. In addition, the condyloid joint between the proximal phalanges and metacarpals allows movement from side to side and at right angles to the plane of flexion and extension. Movements away from an imaginary line drawn through the middle of the middle finger are termed *abduction* and towards it *adduction*. As in all condyloid joints, a very small degree of rotation can also take place; this movement is particularly noticeable in the index finger and forms an important element of opposition movements (see below). The muscles that produce movement at the metacarpo-phalangeal joints are the flexors and extensors of the fingers which have their proximal attachment in the region of the elbow, aided by the small muscles arising within the hand called the *lumbricals* and *interossei*.

There are two systems of flexor muscles (Figs. 20, 21), the superficial and the deep. Passing from the forearm to the hand the muscular portion of the muscle is replaced by a tendon which has the effect of reducing bulk. The tendons forming a circumscribed bunch pass through a narrow tunnel consisting of the carpal bones bridged over by fibrous tissue, and spread out fan-wise so that one deep and one superficial tendon passes into each finger, the deep being behind the superficial.

The deep tendon acts primarily on the last joint of the fingers, while the superficial tendon is responsible for flexing the middle joint. The method by which this seemingly impossible trick is accomplished is beautiful in its simplicity. The tendon of the superficial flexor splits into two, each half passing round the deep tendon with a twist, rejoining behind it and passing distally in this new relationship until it is inserted on the middle phalanx.

The long flexors of the fingers are also adductors, in the sense that not only do they bend the digits but they also produce adduction when the tendons contract. You will notice that on bending the out-stretched fingers the tips approach one another as they fold into the palm. In fact they begin to get in each other's way. The reverse happens (abduction) when the hand opens out. Adduction of the digits (including the thumb) provides the principal mechanism of grasping in non-primate mammals (Fig. 17), and substitutes in some measure for true prehensility. Although primates have evolved a more sophisticated form of grasping using the opposable thumb, adduction (or convergence) on bending and abduction (or divergence) on straightening are movements still retained by the human hand.

An important adjunct to grasping is the extension of the wrist joint. The wrist in 30–40 degrees of extension is in the optimal position for gripping, whether this is mass flexion as in gripping a hammer, or individual flexion as in tapping typewriter keys. In full flexion of the wrist, the hand can exert only 25 per cent of its power in full extension. As is well known, the quickest way to disarm an assailant wielding a knife or a gun is to force his wrist into flexion whereupon he will be forced to drop his weapon, or at least that's the principle!

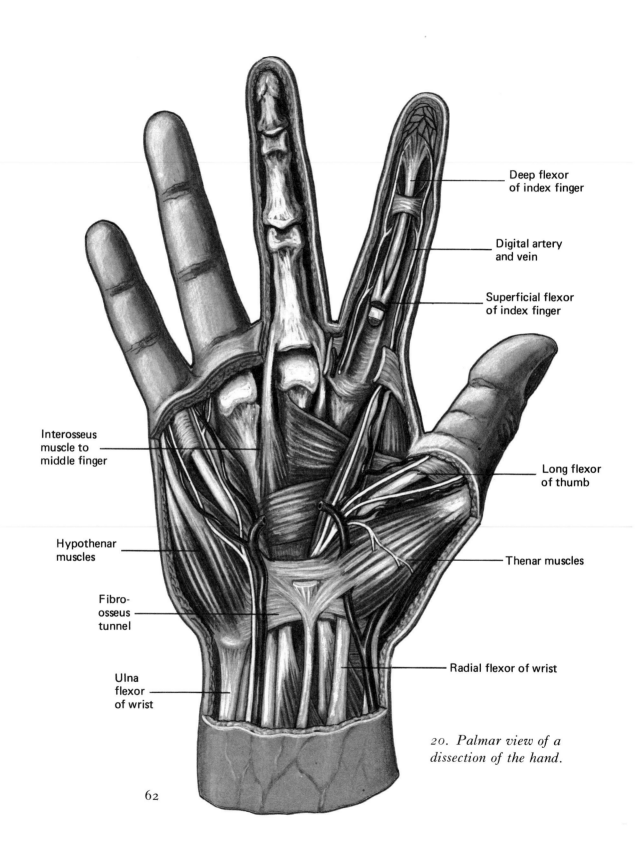

Deep flexor
of index finger

Digital artery
and vein

Superficial flexor
of index finger

Long flexor
of thumb

Thenar muscles

Radial flexor of wrist

Interosseus
muscle to
middle finger

Hypothenar
muscles

Fibro-
osseous
tunnel

Ulna
flexor
of wrist

*20. Palmar view of a
dissection of the hand.*

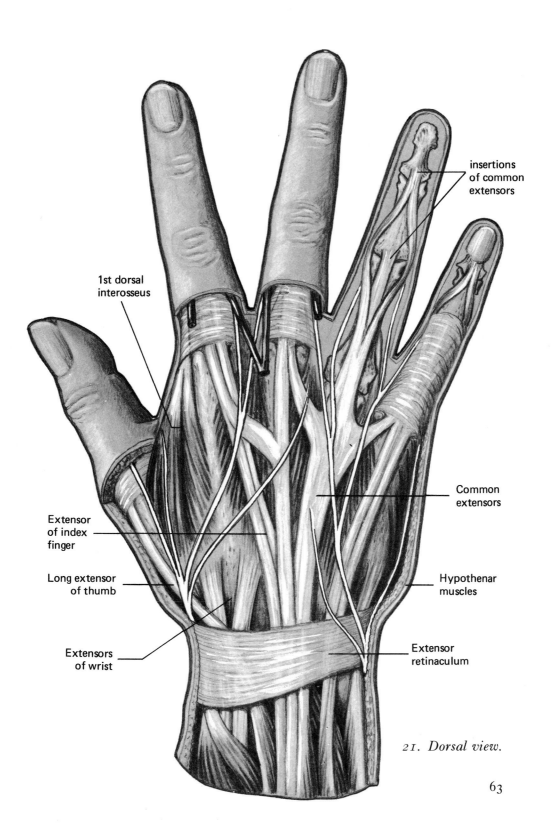

insertions
of common
extensors

1st dorsal
interosseus

Common
extensors

Extensor
of index
finger

Long extensor
of thumb

Hypothenar
muscles

Extensors
of wrist

Extensor
retinaculum

21. Dorsal view.

63

22. Ulnar drift, resulting from imbalance of intrinsic hand muscles following inflammation. A common deformity in elderly people with rheumat‹ arthritis (Courtesy of J. N. Barron).

The long extensor muscles of the fingers differ somewhat from the long flexors at their insertion; to begin with they are not arranged in two layers. Each tendon spreads out into a flattened sheet – what is known as the extensor assembly – over the backs of the fingers (instead of remaining as a rounded tendon, as is the case with the flexors). The index finger in man has two extensors, a superficial and a deep, the rest one only.

Most Old World monkeys such as the macaques have a full set of extra extensors, one for each finger. The orang-utan has a nearly full set of "extra" tendons but gorillas and chimpanzees, like man, possess only an "extra" tendon for the index finger. The index finger is particularly adept in the gorilla, having an almost man-like "aloofness".

Each tendon of the long extensor divides into three slips (Fig. 21): the central slip becomes attached to the proximal phalanx, the other two stream on as a rather undistinguished sheet of fibrous tissue which, once again, forms three slips. The central slip reaches the middle phalanx where it is attached, and the outer two reach the distal phalanx. The tendons of the short muscles of the hand (the interosseus and lumbrical muscles) join the assembly and play an extremely complex role in the mechanism of the hand. These short muscles are extremely well endowed with special nerve endings which provide them with a positional sense that has no equal elsewhere in the body.

Acting independently of the extensor assembly, the lumbricals produce radial deviation of the digits. A paralysis of the lumbrical muscles which occurs following injury to the nerves reaching the hand (see below) accentuates the normal tendency of the fingers towards ulnar deviation. Extreme ulnar deviation – or "ulnar drift" as it is called by clinicians – is one of the most striking features of rheumatoid arthritis (Fig. 22). The deformity comes about as a mechanical result of laxity of the capsules of the knuckle joints due to inflammatory condition of these joints, together with a loss of function of the interosseus and lumbrical muscles for the same reason. Dislocation ulnarwards of the extensor tendons into the gutter between the knuckles follows; the natural tension of the extensor tendons now simply accentuates the normal tendency towards ulnar drift.

Movements of the Thumb

The thumb, the "Lesser hand" as Albinus called it, is the most specialised of the digits. Isaac Newton once remarked that, in the absence of any other proof, the thumb alone would convince him of God's existence.

The generalised mammalian thumb is not a particularly specialised digit and if it has any unique qualities, in a sense they are attributable to its marginal position. Being unsplinted by adjacent digits, it is therefore free to develop a minor degree of extra mobility; the same is true of the little finger.

The primates (with a very few exceptions) show some degree of functional independence of the thumb, although the extent and precision of its movements differs considerably between the members of the Order. One can see a steady advance in the functional complexity of living primates from the lower primates (lemurs and lorises) to the higher primates (monkeys and apes) and finally to man. Monkeys are more adept than lemurs, apes are more adept than most

65

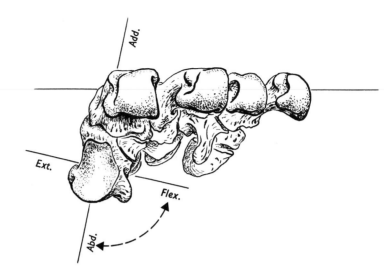

23. The metacarpals are seen end on. The first metacarpal has been removed to show the trapezium. Axes of movements of the thumb are indicated. A combination of abduction and flexion results in opposition. The saddle joint allows a rotational movement indicated by the broken line.

of the monkeys, and man is more proficient still.

The thumb metacarpal is unique. Alone amongst the metacarpals it articulates by means of a freely movable saddle joint with the carpals. The remaining carpals are of the plane joint variety which have very small ranges of movement. The metacarpo-carpal joint of the thumb, being of the saddle type, is almost as mobile as a ball-and-socket joint and has the following movements: adduction-abduction, flexion-extension and medial-lateral rotation (see Fig. 23).

The functional advantage of a saddle-joint is that the two opposing surfaces and their supporting ligaments are so arranged that the stability of the joint is provided without the need for a cuff of bulky muscles disposed around the joint to control and direct its movements, as is the case with other ball-and-socket joints like the shoulder and hip. Bulky muscles at the root of the thumb would seriously impair its manipulative skill and flexibility.

66

Chapter Three
Function of the Hand

Opposition

Perhaps the most important movement of the human hand is opposition. The movement of the thumb underlies all the skilled procedures of which the hand is capable. The hand without a thumb is at worst, nothing but an animated fish-slice, and at best a pair of forceps whose points don't meet properly. Without the thumb, the hand is put back sixty million years in evolutionary terms to a stage when the thumb had no independent movement and was just another digit. One cannot emphasise enough the importance of finger-thumb opposition for the emergence of man from a relatively undistinguished primate background. Through natural selection, it promoted the adoption of the upright posture and bipedal walking, tool-using and tool-making which, in turn, led to enlargement of the brain through a positive feed-back mechanism. In this sense it was probably the single most crucial adaptation in man's evolutionary history.

As it is preferable that opposition should describe the movement of the thumb as a whole, the relative contributions of the three different joints involved are ignored in the following definition:

Opposition is a movement by which the pulp surface of the thumb is placed squarely in contact with – or diametrically opposite to – the terminal pads of one or all of the remaining digits (Fig. 24).

From this definition it is clear that all the digits are involved in the concept of opposition, not merely the thumb. Because it is the most important finger as well as for the sake of descriptive simplicity, the index is used as the opposing digit. Finger-thumb opposition hereafter means index finger-thumb, though much the same arguments apply equally to the other fingers.

The neutral position of the thumb (analogous to a gear-lever "in neutral") is taken as the starting point for a description of the movement of opposition. When the thumb is abducted, it angles from the palm, as the muscles spring into action – into gear, as it were. Once in gear, it moves ulnarwards with a movement of flexion and rotation; the more it flexes, the more it rotates until it comes to a full stop in full opposition. To complete the movement, the index finger flexes at all its joints until it makes contact with the tip of the thumb. Because the thumb is rotated, and because the index finger contributes its own

24. "Perfect" opposition between thumb and index. Note broad area of contact and slightly extended wrist joint.

element of rotation, an intimate contact between the digital pads of the finger and thumb is achieved (Fig. 24). The role of the index finger is relatively passive. Apart from the long flexors, the most important muscle acting on the index finger is the dorsal interosseus, which lies between the thumb and the index finger. It is easy enough to see and feel the interosseus in action when the opposing finger and thumb are tightly pinched together and then relaxed.

Although opposition of the thumb is one of the hallmarks by which man can be authenticated as it were, it is not a movement unique to him. What is unique is the broad area of intimate contact between the finger-tip pulps of the opposing digits that results. The advantage of intimate contact concerns both function and feeling; the greater the surface area of highly sensitive papillary skin available, the more effective is the handling of small and delicate objects.

69

25. Opposition in practice.
For increased stability and
sensitivity the middle finger
is brought into action.

Opposition in Monkeys and Apes

A major factor in achieving pulp-to-pulp contact is the proportionate length of the index finger and thumb. The formula *Total length of thumb* $\times$ *100/Total length of index finger* provides a method of expressing the length of the thumb quantitatively (Opposability Index). A low Index denotes a long first finger and a short thumb; and at the other extreme a high Index has a rather narrow range in primates, with mean values between 40–65. The shortest thumbs and the longest fore-fingers are found in the great apes, particularly the orang-utan (mean: 40). Contrariwise the longest thumbs and shortest fore-fingers are exhibited by man (mean:65). Not all primates are capable of opposing their thumbs. The operative word, you will remember, is *rotation*; without rotation true opposition is not possible. Only the Old World monkeys and apes show such rotation; New World monkeys and prosimians lack this movement. Thus the Opposability Index is applicable only to the Old World primates. New World monkeys and prosimians exhibit what has been called "pseudo-opposability", a movement in which rotation is entirely absent and pulp-to-pulp contact of the thumb and fingertips is wanting.

It is particularly interesting to note that it isn't the apes – as might be expected – whose Opposability Index lies closest to man. The ape

26. Chimpanzee grasping a grape in thumb-index opposition. Note the imperfection of the action due to the presence of a short thumb and a very long index finger. Note, too, the "perfection" of the human opposition.

hand is almost wholly suborned to a specialised locomotor style of brachiation. The apes are brachiators, using their long arms and long hands to swing from branch-to-branch and tree-to-tree. Brachiation pre-empted the function of the hand and relegated its manipulative talents to a secondary (though still important) role (Fig. 26). Gorillas, as has been mentioned, are ground-living animals, seldom climbing trees and never brachiating. Although they have not lost the anatomical hallmarks which identify them as *ci-devant* arm-swingers, they have nevertheless become secondarily adapted to a ground-living way of life. The Opposability Index of gorillas is the highest of all apes, and their manipulative skill with small food objects by far the best developed. Paradoxically, the gorilla, unlike the chimpanzee, is neither a tool-user nor a tool-maker in nature. According to observers in the field, the gorilla totally ignores alien objects planted to test his curiosity and implementation skills.

Much closer to man in manipulative skills, as is suggested by their high Opposability Indices (mean : 57–58), are the baboons and mandrills. Although the hand of a baboon is, in effect, a foot in locomotor terms, it has at least several important functions for which finely adjusted thumb-index opposition is essential. These include feeding on seeds and grain and delicate shoots of young grass which are plucked with finger and thumb; and, grooming. Grooming, or fur cleaning, is both a physiological need and a psycho-social necessity. During grooming, which is performed with great concentration and delicate thumb and index finger work, not only is the skin cleansed of debris and parasites but the social bonds between groomer and groomee are immeasurably strengthened by a service given and received. It is a matter of "I will scratch your back, if you will scratch mine". A relationship based on such trust and co-operation must make for good rapport between members of a troop to the ultimate benefit of the whole species. Tactile communication between male and female through grooming is also an important element in pre-copulatory behaviour.

Grooming is not restricted to the non-human primates; man is an inveterate groomer. It manifests itself principally in verbal communication. One can observe "verbal grooming" in action in any pub, hairdressing establishment, coffee-shop or cocktail party when, for instance, conventional enquiries by the groomer lead the proud parent (the groomee) to discourse at length on the virtues of his/her offspring. Actual physical grooming also has a reassuring and

27. Human (left) and chimpanzee hands in a "position of rest".

tranquillising role to play in the life of man, although such activities lack the concentrated intensity with which they are pursued by monkeys and apes. There can't be much intensity in the classic grooming situation of the wife seeing her husband off to work, straightening his tie and brushing the dandruff off his shoulders.

Movements of the hands as a whole

The concept of the *position of rest*, as proposed by John Hilton (1804–1878) and independently by a Gloucester physician named

73

Thomas Ellis, is no longer talked about nor does it figure in the medical textbooks, yet the principles that Hilton and Ellis propounded are put into practice by surgeons all over the world today. The term – "the position of rest" – was introduced by Thomas Ellis, but the foundations of the concept were laid by Hilton of Guy's Hospital, whose monograph on *Rest and Pain* (1877) is a classic of medical literature.

Thomas Ellis, in his book *The Human Foot* (1889), describes the position of rest as the posture "sooner or later assumed by the limb when in a state of rest from fatigue" or, in other words, during full relaxation or sleep. It is at this time that the outflow of impulses from the brain and spinal cord is much reduced and the muscles are to all intents and purposes fully relaxed. Theoretically the position assumed by the joints is the result of the normal tension of agonist and antagonist groups of muscles in repose, and is such that no part of the ligaments surrounding the joint is on the stretch. The position of rest is the ideal posture (with certain reservations) in which to immobilise an injured joint; it is also the best position should complete stiffness occur, and according to Ellis, in fixing the limb thus we are "doing something towards making complete stiffness the best result to be hoped for". This somewhat equivocal statement is a reflection of the limited opportunities that were available for orthopaedic surgery in Victorian times.

Ellis used to watch the limbs during sleep in order to study the position they adopted and then to follow these "lessons of nature". As a result of his experiments, Ellis concluded that the true position of rest "lies somewhere about a mean between the ranges of easy motion" (Fig. 27). Thus the thumb is in the position of rest at the start of the description of its movement (see above); it lies midway between abduction and adduction, flexion and extension, medial and lateral rotation. The position of the hand at rest has a particular bearing on the rest of this section inasmuch as it lies exactly midway between the activity frameworks of *precision grip* and *power grip*.

There are two classes of movement of which the hand is capable: prehensile and non-prehensile. Prehensile movements are those in which an object, fixed or free, is held by a gripping or pinching action between the digits and the palm. Non-prehensile movements of the whole hand include pushing, lifting, tapping and punching movements of the fingers, such as typewriting or working the stops of a musical instrument.

28. (a) Power and (b) precision grips.

Considering the enormous variety of activities that the hand is called upon to perform, it might be supposed that prehensile movements would be too numerous for simple analysis. However the diversity of movements is more apparent than real; it is not so much that there is a profusion of *actions* concerned in day-to-day activities as that there is a multiplicity of objects involved – switches, doorknobs, latches, cutlery, cups, glasses, pens, pencils, erasers, buttons and coins. In fact, there are only two main patterns and two

75

a

29.　(a) *Lid of screw-top jar is being "started" with a power grip.*
　　(b) *Once it is loose the grip changes to one of precision.*

b

76

subsidiary prehensile patterns. The main ones are the *precision grip* (Fig. 28b) and the *power grip* (Fig. 28a). The subsidiary patterns are the *hook grip* and the *scissor grip*.

PRECISION GRIP is executed between the terminal digital pad of the opposed thumb and the pads of the fingertips. Large objects held in this way involve all the digits, but smaller objects require only the thumb, the index and the middle digits. The precision grip is employed when delicacy of handling and accuracy of instrumentation are essential and power is a secondary consideration.

POWER GRIP is executed between the surface of the fingers and the palm with the thumb acting as a buttressing and reinforcing agent. Under some conditions the thumb supplies directional control. Thus, there is a certain element of precision even in the power grip, just as there is an element of power in the precision grip. With heavy tools such as coal-hammers which need little precision in their use, the thumb reverts to its crude reinforcing function, wrapping over the backs of the digits forming a fist). In making a fist, power is the only consideration and accuracy does not come into it.

HOOK GRIP is a subsidiary grip that is a function of the flexors of the fingers; the knuckle-joints are straight and the two terminal joints acutely bent. The hook grip is used when carrying a heavy suitcase, when opening a sash window from the top, or when grasping specialised tools such as pliers or wire-cutters. Although the hook grip is a prehensile posture, it is not a very important one in the general functioning of the hand.

SCISSOR GRIP is a subsidiary grip pattern in which an object is grasped between the sides of the terminal phalanges of the adjacent index and middle fingers. Man may grip a cigarette in this fashion and occasionally he uses it casually to pick up small, flat objects when he cannot bring his thumb into action.

It is important to appreciate that the type of grip used in any given activity is *a function of the activity itself and does not depend on the shape or size of the object gripped*. For example, an electric light bulb is held in a precision grip when its lugs are being engaged in the socket (Fig. 30) and in a power grip when the bulb is being screwed or locked

77

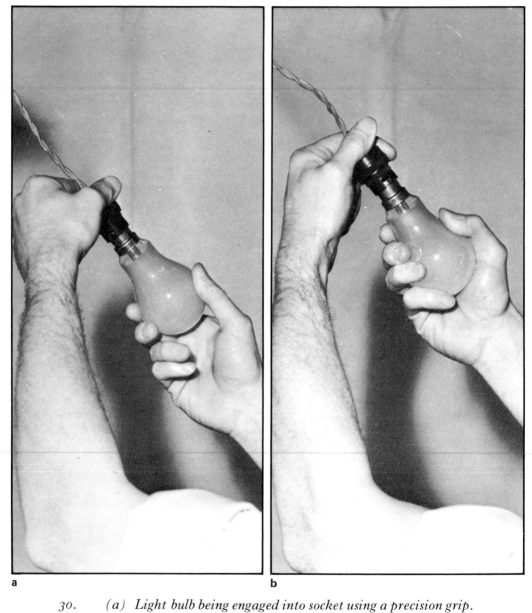

a   b

30.   (a) Light bulb being engaged into socket using a precision grip.
*   (b) Posture of hand changes to one of power grip in order to lock*
*   it home.*

home (Fig. 30b). A screw-top jar demands a similar sequence of grips (Fig. 29a and b). The lid is held in a precision grip when the threads are being engaged, but the grip changes to one of power when the lid needs to be screwed home tightly. In these examples the object being manipulated (the light bulb or the lid) remains the same throughout the sequence; it is the nature of the activity that is changing.

If the object gripped is very small or very large, the generalisation may not hold, however. A needle is too small for a power grip and a coal-hammer too large and heavy for a precision grip. In these circumstances there is no choice. The grip may also be affected by the shape of a custom-built tool or handle; for example, a pistol-grip handle with indentations for the digits cannot easily be used in a precision grip.

Precision and power grips are functional concepts, but they are to some extent discrete as far as their nerve supply is concerned. The brunt of a paralysis affecting the median nerve falls on the muscles responsible for the precision grip, so this is the "nerve of precision". The ulnar nerve supplies the bulk of the power grip to muscles and can be referred to as the "nerve of power".

Prehensile Patterns in Primates

Precision and power grips in their fullest expression are hallmarks of mankind, but the evolutionary trend towards the establishment of a dual prehensile pattern takes place at a relatively primitive stage of primate evolution as represented by the New World monkeys. Prosimians (lemurs and lorises) possess only a single prehensile pattern; the hand opens and closes like the jaws of a toy crane in an amusement arcade (Fig. 31). There are no really precise movements and no power movements as such – merely all-purpose grabs. New World monkeys are not much better equipped, although workers who had made a particular study of South American species claim the two grips are distinguishable, particularly in the more advanced species, such as the capuchin monkey. The precision grip is absent, of course, in the most specialised of the New World monkeys – the acrobatic spider monkeys which are thumbless. Spider monkeys can achieve a modicum of precision by using the scissor grip, in which the object is grasped between the sides of the second and third digits. The prehensile tail of these monkeys is both a power and precision organ. The tail can be used to suspend the body from an overhead branch (power grip), or to delicately pick up such small items as a shelled peanut (precision grip). The undersurface of the lower one-third of the tail is covered by naked skin, identical in function to the papillary skin of the finger.

Old World monkeys are fully differentiated as far as their prehensile grips are concerned, the only exceptions being the colobus monkeys which, like the spider monkeys, have no thumb at all. Unlike the latter, though, the colobus' tail has no compensatory prehensile powers.

Apes, the gorilla, orang-utan and the chimpanzee, show different-iation of the two grips quite clearly although the precision grip – for mechanical reasons – is less adept than in some monkeys. Historically, the apes are brachiators and towards this end the hand has undergone a reduction in thumb length and an increase in finger length. Thus finger-thumb opposition (the basis of the precision

31. A galago about to pounce on a mealworm. Note that all digits are prepared for action.

grip) is considerably impaired (see Opposability Index, p. 70); occasionally the scissor grip is used.

During brachiation, the great apes make good use of their elongated fingers which act as a suspensory hook. The power grip in chimpanzees differs somewhat from man's; the object being gripped is orientated transversely rather than obliquely in keeping with the relatively uncomplicated hinge pattern of the metacarpophalangeal joints (see Fig. 27). This grip is also reflected in the transverse orientation of the simian crease. As we have noted, to retain a "hook grip" is supremely important to apes but is only marginally so to man.

Prehensile Patterns and Tool Handle Design

An extension of the two-grip concept is its application to the design of tool-handles. Most tools and appliances have two ends – the working-end and the handle. The form of both components should be dictated by the nature of the particular operation involved, but too frequently it seems that inventiveness and ingenuity is concentrated on the working-end, and the handle-end is left to take care of itself. The human hand is highly adaptable and can secure a grip on almost any kind of shape, but if the tool is to be of maximum efficiency, the handle must be designed for a specific function.

It should be a relatively simple matter to design a handle for a single-purpose tool. A careful analysis of the intended activity under all sorts of environmental conditions and situations (cold and hot weather, greasy hands, overhead working, etc.) will determine the most effective grip, whether precision or power. However errors of judgement are still only too common. Let us take the types of nailbrush as examples. One model on the market recently was ball-shaped and made of a shiny plastic material. Not only was the basic shape unsuitable for a power grip (which is the most effective grip to be used when scrubbing the nails) but the rounded, smooth surface, wet with soapy water was impossible to grip at all. A common contemporary design is the rectangular brush surmounted by a half-hoop in which the fingers are supposed to be inserted; seldom is the hoop wide enough to admit the fingers, and never does it permit the normal staggered flexion of the fingers of a power grip.

Many hand-drills aimed for the domestic market are equipped with a small knob on the driving wheel which is too small to be held in a power grip; a precision grip (quite unsuitable in the circumstances)

is all that can be achieved. By means of a simple experiment, I have been able to show that increase in the diameter of the knob until it reaches the dimensions that can comfortably be held in a power grip substantially improves the efficiency of hand drilling.

The handles of drills and similar appliances are seldom "set" in the ideal functional position for the hand and forearm; as a result such postural corrections as are necessary can only be made by adjustments of the muscles of the shoulder and trunk, which are physically tiring. Fig. 32 shows the hypothetical stages in the design of a tool for the domestic market. There is nothing to go wrong at its working end – sandpaper is sandpaper – but there are plenty of errors that can creep in in the designing of its handle.

Man has passed from tool-user to tool-maker (see p. 126) and now – somewhat ironically – back to tool-using again. There are a few craftsmen left who make their own tools, but not many; tools are no longer personal creations of craftsmen (whose survival once depended on their effectiveness) but are the standardised products of commercial tool-makers. Tools made for the uncritical domestic market are the worst offenders; shapes of handles are more often chosen for their packaging and their modern design qualities than for their functional suitability. Early man in the throes of constructing a hand-axe never made that mistake; his life and livelihood depended on it.

32. Theoretical stages in the design and development of a sander for which a power grip is required:
  (a) The first model
  (b) Rejected because of inadequate space for fingers
  (c) Rounded knob substituted
  (d) Rejected since knob provides inadequate surface for power gripping
  (e) Handle substituted
  (f) An improvement, but fingers cannot be wrapped round cylinder sufficiently intimately to exert full power
  (g) Handle raised and set obliquely in power gripping position
  (h) Final product satisfactory.

Chapter Four
Evolution of the Hand

Evolution of the Hand

Man and primates live what might be called a "hand-to-mouth" existence. In contrast to most non-primate mammals who put their mouths to the food, primates carry food to their mouths. As might be expected, such a fundamental difference in behaviour is accompanied by some notable anatomical modifications, the most outstanding of which are prehensility of the hand, shortening of the snout and migration of the eyes to the front of the face.

The evolutionary predecessors of primates, the insectivores, constituting shrews and their kind, were long-snouted, nocturnal creatures with high-cusped puncturing and crushing teeth. The sense of smell dominated their way of life; insectivores navigate with their noses. Noses have a tactile as well as olfactory role to play in searching for food. The snout is used to locate and identify the prey, the jaws to seize it and the teeth to pierce and kill it. Insectivores with their long snouts and highly developed sense of smell might be described as "noses on legs". Migration of the eyes to the front of the face, along with refinements in the eye itself and in the visual pathway to the brain, shifted the sensory emphasis towards vision and away from smell. The earliest true primates were vision-orientated and therefore – one can assume – diurnal in habit. All this was happening some 65 million years ago.

Primates share their hand-to-mouth feeding habits with a number of mammals which are exceptions to the general rule. Squirrels and other rodents, bears, raccoons and otters are also hand-feeders, but there are behavioural differences that set the primates apart. For example, marmots and squirrels feed by holding food objects in *both* hands rather than in one (see Fig. 17). The combination of two non-prehensile hands substitutes for the single-handed prehensility of primates.

The matter of the giant panda's notorious "thumb" has already been mentioned (p. 30). The discovery that the panda feeds itself one-handed would seem to contradict the principle that non-primate mammals are two-handed feeders, but once again anatomy comes to the rescue of behaviour; an enlarged wrist bone mimicking the thumb is a unique adaptation, a one-off event, that does not really affect the general principle.

Arthur, the famous cat-food cat of a British TV commercial, appears to be another exception to the two-handed rule, since day after day he can be seen scooping fish out of a can with one paw. However, fish is soft and Arthur's claws are sharp. I submit that

Arthur is really a red herring.

Eating has been a prolific fount of social rules throughout the ages, the eighteenth and nineteenth centuries in particular having much to answer for. Which part of the spoon should be used for supping soup? Should bread be cut or broken at the evening meal? Is it incorrect to hold your knife as if it were a pen? Which way should the port circulate?

Table manners and conventions, in maintaining the unity and cohesiveness of the group, serve a very useful social purpose, analagous to the grooming behaviour of baboons and macaque monkeys. Manners represent a symbolic reiteration of the current rules of social class and, at the same time, they represent a cultural discipline which insists that the conventions are adhered to by gentlemen even if, like Charlie Chaplin, they are snowbound in the Yukon with only the toughest of leather soles to stave off the pangs of hunger. Carruthers, too, always dressed for dinner in the most putrid reaches of darkest Africa, in spite of incessant drums and restless natives.

However, table manners are not innate but learned and can be temporarily suspended by group decision as was the case, in a genteel sort of way, with Marie Antoinette and her court. Less histrionic than life at Le Petit Trianon is the picnic where even the most fastidious will gnaw on the leg of a chicken, Charles Laughton fashion, and fling the bones over the shoulder into the nearest bush. To eat alfresco provides the welcome excuse for temporarily rejecting the social rule that food should be eaten with a knife and fork. Man is a hand-to-mouth feeder just like the rest of the primates, but in Western civilisations convention insists that we try to conceal the fact.

Classification of Primate Hands

A monkey's paw has a dual role; it is required to act as both hand and foot. In monkeys a compromise has been reached by which both functions can be served. Man's hand has only a single role – that of prehension – and so no adaptive compromise is required and natural selection has had the easier task.

The Foot-Hand
Man alone among the primates has fully emancipated his hand from

87

any locomotor role once his childhood has passed. When young, he crawls quadrupedally with the flat of his hand applied to the ground like a monkey. He does not, as might be expected from his immediate ancestry, walk on his knuckles as an ape does (Fig. 33). As he grows up, the human child passes, briefly, through the foot-hand phase that a monkey enjoys throughout life.

The foot-hand of a monkey, as its name suggests, is capable of both functions. The presence of prominent palm pads, the cushion of palmar skin on the ulnar side and its extension on to the wrist, forming a sort of "heel" to the hand, provides evidence of locomotor use, while the flattened nails, expanded digital pads, and the mobile thumb bear witness of manipulatory capability. Both sets of characters are equally well developed, with the result that Old World monkeys combine expert handling of objects with efficient locomotion. Naturally, there are limitations to the possession of a foot-hand, otherwise there could certainly have been no selection in man for bipedalism. The monkey's foot-hand is prevented by locomotor commitments from carrying objects from place to place, while bipedalism leaves the hands free for this purpose. The carriage of food was probably the most significant benefit accruing from the bipedal gait, but the carriage of other objects, such as weapons, tools and infants, would also have provided undoubted evolutionary advantages for the emerging humans.

33. Orang-utan and human infant. Note that the former knuckle-walks while the human being crawls in a plantigrade fashion like any monkey.

The True Hand

Primates with true hands are the apes and man. Although superficially unalike, the ape and human hand are cast from the same mould. In contrast to the foot-hand of monkeys, the "heel" of the *true hand* is wanting, a clear indication of the absence of weight-bearing on the flat of the hand. The palmar pads have also disappeared as circumscribed prominent hillocks, although the remnants still persist in true hands as reminders of the past. Dissimilarities between ape and human hands are largely quantitative; the ape hand is elongated in its metacarpal and phalangeal segments and the thumb is very short; this is particularly true of the orang-utan. The gorilla has the longest thumb and the shortest fingers of all the great apes, and is thereby closest to man in its manual proportions.

While apes never use the hand as a foot on the ground (although the reverse is true), it plays a major part in brachiation. Nowadays the African apes are rather poor brachiators. Whatever they did in ancient times, they are now found more often on the ground than in the trees; in fact the male gorilla *never* leaves the ground if he can help it. Locomotion for the gorilla and the chimpanzee is a bit of a problem. Anatomically they are not adapted for quadrupedalism nor are they equipped to walk bipedally for more than a few steps at a time. The solution is a sort of compromise, an expedient called *knuckle-walking* in which the weight of the forepart of the body is borne on the knuckles (see Fig. 7); thus, since the arms are longer than the legs, the body is supported in a semi-upright posture. In spite of bearing weight, the hand is not acting as a foot but as a hand; gorillas and chimpanzees, in fact, walk on their hands. Knuckle-walking is a well-established gait associated with a number of special characters of the hand and forearm that make it possible. On occasion man uses his hands to suspend himself in much the same way as brachiators do, although his size and his weight-to-muscle ratio militates against it except in the young (shades of those horizontal ladders in the school gym!) or in specially trained acrobats. Furthermore, in special circumstances man uses a knuckle-walking posture, not to walk but to support his body upright, for example, when standing at a table, delivering a speech.

To sum up, the great majority of primates possesses a *foot-hand* which is fitted for both manipulation and locomotion. A minority possesses a *true hand*, which has a dual purpose in apes, but a single purpose only, manipulation, in man.

GEOLOGICAL TIME SCALE
TERTIARY AND QUATERNARY

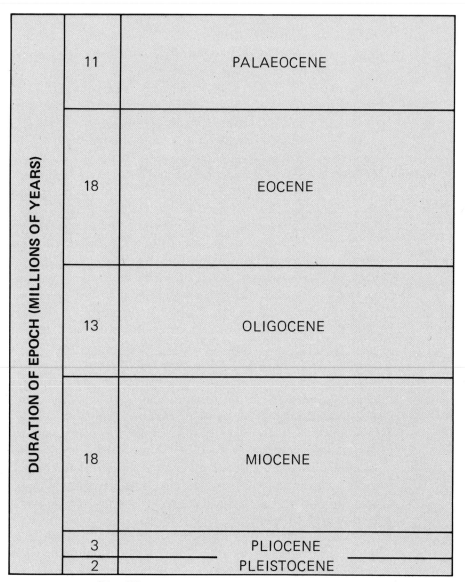

DURATION OF EPOCH (MILLIONS OF YEARS)		
11	PALAEOCENE
18	EOCENE
13	OLIGOCENE
18	MIOCENE
3	PLIOCENE
2	PLEISTOCENE

65 million years – approximate total duration

34. Geological epochs during the Tertiary Period.

Origins of Prehensility

For nearly 100 years the question of whether mammals were derived from an arboreal or terrestrial ancestral stock has been a key one among biologists. While it is unarguable that all modern primates were derived from an arboreal stock, there is no convincing evidence that the ancestral mammals were.

All primates possess divergent and opposable (in its widest sense) big toes and the question is this: are prehensile big toes an ancient mammalian heritage dating back to the Jurassic and Lower Cretaceous when the first true mammals evolved, or are they a relatively recent adaptation to life in the trees? If the fossil feet of ancestral placentals were known well enough the puzzle might be solved, but the truth is that there is no adequate evidence to show if placentals possessed divergent and grasping big toes before the Eocene (see Fig. 34).

The palaeontological basis for arguing an arboreal ancestry is that *Didelphis*, an opossum-like form, possessed a divergent big toe, but *Didelphis* is a marsupial. Since there is no suggestion that the marsupial grade was a staging-post for all primitive mammals on their way to becoming placentals, the existence of prehensility of the foot in early marsupials is really quite irrelevant. Arboreal the earliest marsupials may have been, but the placental stock as far as we know was ground-dwelling, living a nocturnal insect-eating life on and under the forest floor among the roots of the trees. An arboreal background for placentals remains only a theoretical possibility and we must look elsewhere for the origins of prehensility.

A divergent big toe or thumb is usually associated with the power of prehensility, the ability to seize or grasp an object and hold it securely in one hand; but there is more than one way of achieving this. For example, neither the spider monkey nor the colobus monkey have thumbs, yet both are capable of effective grasping. The basis of their prehensility is finger flexion. Having exceptionally long fingers relative to the rest of the hand and by folding these digits into the palm, a powerful grasp can be achieved. The chimpanzee and the orang-utan, with a technique of grasping slender supports by means of the double-locking mechanism (see Fig. 11), have remarkably short thumbs and excessively long palms, a situation that approaches the colobus hand in its functional proclivities. Even clawed mammals with long digits can achieve a measure of prehensility. The mar-

35. The marmoset hand. Note palmar pads, narrow finger tips and claws, and little differentiation of thumb. This is a primitive form of foot-hand.

mosets which have short clawed digits and a long narrow palm provide a good example (Fig. 35).

Increase in size and a shift in diet appear to be two of the major factors involved in the evolution of prehensility in the primates. The effect of a change from insect-eating to fruit- and leaf-eating is that a different part of the trees is used for feeding.

Broad-leaved trees, particularly those with spreading or umbrella-shaped crowns, are characteristically hollow within. As Defoe said in another context: "Crowns are empty things". Fruit and leaves are found in their greatest abundance on the slender, whippy branches at the extreme periphery of the crown. This small-branch setting is the feeding zone for the leaf, fruit and bud eaters, and it was this milieu that must have attracted the earliest arboreal primates.

Provided that the ratio of branch size to body size remains high, the axis of gravity is perpendicular to the axis of the branch and the centre of gravity is low, arboreal animals can maintain their balance on the branches of trees by means of claws or digital pads alone (Fig. 36a). Locomotion on such branches of trees offers no particular problems since, for small animals, branches provide broad aerial highways theoretically no different from comparable substrates on the ground. With increase in body size a critical point is reached where a potentially unstable system is produced and lateral displacement of

92

the axis of gravity will lead to overbalancing (Fig. 36b). Without some compensatory trick such as prehensility (Fig. 36c), large animals would find it extremely difficult to move about among slender, leaf- and fruit-bearing branches and keep their balance; thus, a rich food source would be closed to them. The rule must have been: stay small or if you must get bigger, then evolve prehensility at the same time!

A talent to suspend the body while feeding in trees is a characteristic of most primates, though not used by all to the same degree. The adaptive value of suspension is that the feeding range, theoretically, is thereby doubled to include leaves and fruit above and below the branch (Fig. 37b).

The suspended posture may be quadrupedal, bipedal or bimanual. Bimanual suspension is of particular interest since it is practised by

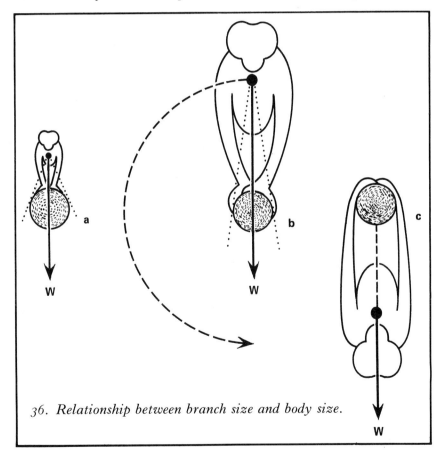

36. *Relationship between branch size and body size.*

the gibbon and occasionally by the rest of the man-like apes. It is also a familiar sight to see a gibbon hanging by one arm, the body twisting and pivoting as the long free arm reaches out and around to pluck the fruit.

The South American spider monkeys, woolly monkeys and howler monkeys are also habitual "suspender-eaters", supplementing their prehensile limbs with a prehensile tail.

The feeding behaviour of a quadrupedal monkey, such as the macaques, the langurs and the guenons, is to sit precariously on slender branches, well anchored by their grasping feet, and bend back the terminal fruit- and leaf-bearing twigs so that they may be plucked by hand (Fig. 37a). The emphasis in these quadrupeds has not been so much on suspensory-prehension as on supportive-prehension; such a pattern of behaviour does not lead to the selection of long arms, hook-like hands, and modifications in the shoulder region that took place among the arm suspenders. Arm-suspension and arm-swinging have evolved many times in parallel among the primates; particularly striking is the extraordinary morphological similarity between spider monkeys of the New World and gibbons of the Old, both suspender-eaters but otherwise only remotely related. Parallelism, as this evolutionary phenomenon is called, has played a major part in the evolution of the primates.

Thus a dietary change may well have been the initiating factor in the evolution of prehensility, a plausible chain of events that can be summarised as follows. Assuming that the habitat of the earliest primate insect-eaters was the deep litter and low shrubs of the forest floor, the first effect of a change of diet would eventually have been a change of habitat by exploitation of the trees where food was more prolific and of a different sort. Feeding in trees on fruits and leaves would have raised many problems concerned with life high above ground, not the least of which would have been the question of stability. Claws would have been reduced to nails and digital touch pads on the tips of the fingers would have become important. Divergence of the big toe probably preceded divergence of the thumb, but soon competition would have placed strong selection pressures on the increased mobility of the thumb and the manipulative capacity of sensitive touch pads. Thus, the forefoot would have become a foot-hand.

Opposition is a much misused word and has been employed freely to label any movement of the thumb whether or not *rotation* (which is

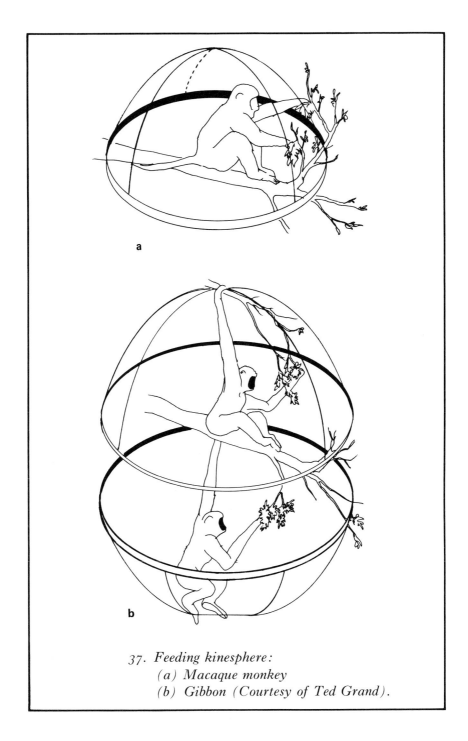

37. *Feeding kinesphere:*
    (a) Macaque monkey
    (b) Gibbon (Courtesy of Ted Grand).

the key movement, see p. 68) is occurring. For example, to describe the movement in New World monkeys as opposition would be quite incorrect, for rotation of the thumb is absent. Since, superficially, the movement of the New World thumb is broadly similar to thumbs which actually do show rotation, the term *pseudo-opposability* is used.

Fossil Hands

The earliest scrap of fossil evidence about primates comes from a place called Purgatory Hill in Montana, USA. It tells us nothing about the hand; in fact it tells us very little about anything beyond the fact that the sometime owner was possibly a primate. The scrap in question was an upper molar tooth, and it has been assigned to a new genus called *Purgatorius*, a genus which was living contemporaneously with the last of the dinosaurs, 65 million years ago. Even to some case-hardened palaeontologists, the assignment of *Purgatorius* to the primates on the basis of a single tooth seems presumptuous; it may indeed have been ancestral to monkeys and apes, but there is the distinct possibility that its primate-like shape was merely a functional adaptation to a particular diet and way of life.

The fossil family, the Plesiadapidae, found in North America and Europe, provide much more information about early primate life. They came in a variety of different sizes from rat-size to cat-size; they were ground-living, probably diurnal and were fruit- and leaf-eaters. Their feet were essentially feet with no specialisation of the thumb; on the end of each digit were strong claws. It was possible that some species had adopted arboreal life but had not yet undergone any arboreal adaptations. At this stage – some 60 million years ago – there is no evidence of development of the typical primate foot-hand with its functional traits of prehensility and opposability.

Non-arboreal primates, such as the Plesiadapidae, were quite likely to have come up against the strong selection pressures of rodent competition in the ground-living niche which they both shared. For many years it was a popular, if somewhat simplistic, theory of primate arboreality that the evolving rodents in effect pushed the primates up into the trees by pre-empting the available niches at ground level. I am satisfied that this is one reason at least why primates became arboreal, though not the sole one. A preference for edible tree products (including insects) was undoubtedly another compelling factor; an insectivorous element figures in all primate diets.

The Plesiadapidae were the last of the early ground-living primates. From that time on primate evolution took place in the trees, the only environment in which the major distinctions of the primate hand could have developed. Paradoxically, certain refinements of the human hand evolved in quite another milieu, but more of that later.

By the middle Eocene (see Fig. 34), an important group called the Adapidae were flourishing in North America and Europe. Two American genera, *Smilodectes* and *Notharctus*, are well known from fairly complete fossil skeletons. The adapids were rather like modern-day lemurs, with hindlegs rather longer than forelegs and slim acrobatic bodies. For the first time in the fossil record we find in creatures of an arboreal habitat that not only are the digits surmounted by nails rather than claws, but the hands are prehensile. The fingers are long and the thumb is mobile, probably pseudo-opposable.

Towards the end of the Eocene, a group of primates emerged whose descendants, the tarsiers, are still with us today in southeast Asia and the Philippines. The extraordinary thing is that the Eocene tarsier and the modern tarsier are virtually the same animal, made up of the same components. We can safely assume that the hands were identical with their modern counterparts. Instead of two distinct prehensile patterns found in monkeys and apes, the tarsier has only one pattern which is neither a precision grip nor a power grip; for want of a better word, and at the same time to illustrate the catch-as-catch-can nature of the movement, we can call it a grab. Galagoes show much the same grabbing movement (see Fig. 31), and indeed it is likely that by the end of the Eocene no precision or power grips as defined had yet evolved in the primate stock.

The Oligocene was the geological epoch when the monkeys and apes first appeared (23–26 million years ago). The site of the most important fossil remains is a region called the Fayum in Egypt, which was at one time an area of lush tropical forests well supplied with rivers and swamps, now an arid waste of desert sand. Although skulls and jaws have been found there, very few limb bones have turned up, so this epoch is a blank as far as hand evolution is concerned. However, working forwards from the Eocene and backwards from the Miocene (when hand bones are known) we can build up a picture of a hand with fairly long fingers, a pseudo-opposable thumb,

38. The hand of Proconsul africanus, *a fossil pongid from Kenya, probably related to the modern chimpanzee, partly reconstructed. The proportions of the hand are not chimpanzee-like; the thumb is not truly opposable. 19 million years BP (before present).*

98

sensitive touch pads on the fingertips and the beginnings of power and precision grips, much like the hand of New World monkeys today.

The first unequivocal ape ancestor that we know of lived 19 million years ago, was called *Proconsul* and was about the size of a fox-terrier. *Proconsul* fills a very important gap in the fossil record because it lived at a time when the "great divide" took place – the dichotomy between apes on the one hand and men on the other. In fact to many people it represents the common ancestor, the progenitor of the ape-human stock. *Proconsul*, discovered in East Africa in 1948 by Dr and Mrs L. S. B. Leakey, was undoubtedly an ape (not a monkey) whose teeth were very similar to a living chimpanzee's. Surprisingly, below the neck it was more monkey-like than ape-like; unlike apes, *Proconsul* was not a brachiator but a fairly typical quadruped who ran, climbed and leaped among the trees, but did little arm-swinging. The characteristic gait of apes evolved later and – as already pointed out – led to a rather specialised sort of hand. Man, who was never a true brachiator, though he may have done some arm-swinging in his evolutionary youth, at the common ancestor stage, perhaps, avoided the pitfalls of the specialised brachiator's hand.

Fortunately a single specimen of the left hand of *Proconsul* was discovered in 1950 in an infilled pothole in the floor of an ancient buried lake; most of the arm bones were found with it (Fig. 38).

The hand was flat-nailed, prehensile and with a pseudo-opposable thumb. The general shape was a cross between a monkey's hand and a chimpanzee's. The metacarpal heads were smooth and globular and totally lacked the sesamoid gutters which are so characteristic of Old World monkey hands; the carpal bones were ape-like in their shape and function except for the trapezium, the bone that articulates with the base of the thumb. Instead of a saddle-shaped surface denoting true opposition, the joint was cylinder-shaped (as in modern New World monkeys), indicating pseudo-opposition. *Proconsul's* hand was monkey-like in overall proportions inasmuch as the fingers were rather short and the thumb rather long, the proportions being the reverse of the chimpanzee's hand.

One must assume that when the descendants of *Proconsul* turned their backs on an arboreal life and started the long evolutionary trek to manhood through life on the ground, the hand was still a foot-hand. Only with the assumption of an upright posture and a bipedal walking gait would natural selection have favoured the hand as a non-weight bearing multifactorial instrument, gradually elminating the characteristics of the foot-hand (the palmar pads and "heel" of the hand for example) and replacing them with expanded finger tips, and a stout and truly opposable thumb and a broad palm.

Unfortunately from the time of *Proconsul* to the era of the australopithecines, the South Africa near-men, a time interval of perhaps 14 million years, we have no record of the hands. This is particularly disappointing in the case of *Ramapithecus* (10–14 million years old) a widespread genus (Africa, India and Europe) of what most people would accept provisionally as the earliest fossil known in the human line. We have plenty of jaw and dental material of *Ramapithecus* (which is itself very informative) but not a single bone below the neck. Once again anthropologists have to resort to the educated guess which is that *Ramapithecus* was bipedal which left the hands free for the carriage of weapons. There is no convincing evidence of tool-making and therefore no way of telling whether or not this human forerunner utilised a precision grip.

The hands of the South African australopithecines are known from a few bones the most instructive of which was a thumb metacarpal complete and undamaged from Swartkrans in the Transvaal. There was a good saddle-joint at the base of the bone and one can take it that the thumb was opposable. The bone itself was exceptionally robust and the strong muscular markings indicated powerful action of the thumb. In contrast, the parts of two other metacarpals that are known are finger metacarpals, slenderly built and devoid of muscular markings. It seems most likely that the thumb metacarpal belonged to one species of australopithecines (*Australopithecus robustus*) and the finger metacarpals to the other (*Australopithecus africanus*). The former species, a gorilla-like vegetarian, was physically robust, while the latter was a more delicately boned, gracile creature who was a hunter of sorts and certainly a tool-user, if not a tool-maker. Both species were bipedal.

In 1960 at Olduvai Gorge, Tanzania, Louis and Mary Leakey made an important discovery. A number of hand and foot bones, two skull fragments, a juvenile jaw bone and a collar-bone were found in Bed 1 (the lowest of the strata exposed on the face of the gorge) associated with a well-defined "living floor" including tools of a simple form. It was this latter material which led to naming the fossil *Homo habilis*, the "handyman".

The Olduvai hand assemblage originally consisted of twenty-one bones, and was first described by the author in 1962. Seven of the bones turned out to belong to some non-human species, and one was not a hand bone at all. Of the remaining thirteen, two were adult and eleven were either juvenile or indeterminate (probably juvenile) in age. Most of the juvenile bones belonged to the right side.

The hand is broad and short and quite robust considering the age of the individual. The bones lack the slenderness of a chimpanzee or an orang-utan; nor do they possess the sheer craggyness of the adult gorilla hand. Morphologically the Olduvai hand cannot be closely matched with any human or ape species living today; its nearest counterpart would be a combination of features of adult man and a young gorilla (Fig. 39).

The most striking human features are the breadth and power of the terminal phalanges, particularly of the thumb which clearly carried a broad flat nail. In combination with a very well developed saddle joint, the thumb gave the appearance of a notably strong digit.

Functionally, it is very probable that the power grip was well developed and effective but there is some doubt about precision grip which, while undoubtedly possible, may not have been as fully evolved as in modern man.

The discovery of *Homo habilis* at Olduvai Gorge revolutionised our preconceived ideas of the time-scale of human evolution. During the fifties we scoffed at the idea that man was as much as a million years old, but by 1961 we were in possession of the astonishing fact that, indeed, he was nearly twice as old, 1·75 million in fact (*Homo habilis*). Richard Leakey, in 1973, produced good evidence that the skull with the site number KNM–ER 1470 from Lake Turkana, Kenya, was not only clearly *Homo* but bore a potassium-argon date of approximately 2·5 million years BP.

Of the many remarkable discoveries in Africa during the last decade the rich fossil deposits of the Afar depression in East Central Ethiopia, inland along the banks of the River Awash, promise to be the most sensational because it seems, if preliminary diagnosis is correct, that true men of the genus *Homo* were occupying this area over 3 million years ago. Now, if Dr Johanson and his colleague, Dr Maurice Taieb, the discoverers of the human material at Afar, are correct, the Human Club has to stretch a point and admit the membership of some very ancient men indeed – at least 3 million years ancient (Fig. 40).

At least two genera of early man appear to be present in the Afar

2 cm

39. The hand of Homo habilis *found at Olduvai Gorge, Tanzania, in 1960. Assessed at being 1.75 million years BP. Remains of hand found at the same level as simple stone tools called pebble choppers. Thumb fully opposable (Courtesy of M. H. Day and J. L. Scheuer).*

40. Fossil hands from Afar, Ethiopia. 3 million years BP (Courtesy of Michael E. Bush, Cleveland Museum of Natural History).

formations – true man and an australopithecine or near-man; fortunately for our special interests, hand bones of both have been recovered, though not yet officially reported upon. The material consists of a right capitate, hamate, and trapezium, a left pisiform, 18 metacarpals and 37 phalanges. When this material is sorted out and studied, we may be in for a few surprises about the evolutionary history of the hand.

Apart from a single wrist bone of Pekin Man (*Homo erectus*, three-quarters of a million years old), no hand bones are known from the time of *Homo habilis* at Olduvai Gorge 1·75 million years ago until the age of Neanderthal man of Europe, the ice-beleagured side-branch of the human race, 50,000 years ago. The villains of the piece largely responsible for this shortfall, and the worst enemies of palaeontologists, are the animal scavengers which eat dead flesh, dismantle the limbs and crunch the bones; acid soils and bacteria doing the rest. It's a wonder that palaeontologists have anything to work on at all.

Mercifully for our knowledge of the postcranial remains of Neanderthal man, the custom of burying the dead evolved during Mousterian times. Cave-living, too, probably offered some protection from scavengers. Whatever is the explanation, Neanderthal hands are quite well known though few actual studies have been made. One such, using the methodology of multivariate analysis, was able to show that the hand bones had certain unique anatomical features. Jonathan Musgrave was also able to show that the degree of dexterity in Neanderthal man was probably not as high as in modern man.

Neanderthal man is everyone's idea of a human ancestor (though he was almost certainly *not*, being a side branch and not ancestral to modern man). He is the original caveman of popular imagination, with a squat muscular body and a brutish face. For good measure he carried a knotted club, fought off attacks by dinosaurs and dressed his women in furry bikinis.

In fact, anatomically speaking, he was a good, large-brained *sapiens* type differing only slightly from modern man. It must not be forgotten that Neanderthal man, during the glacial, was living effectively within the Arctic circle and it is to this ecological fact that he owes many of his physical novelties. Neanderthalers, in fact, were a very spectacular race of "eskimos" with short, sturdy bodies for conserving heat, large knobbly joints and massive jaws that could chew tough, partially cooked food.

41. *Fossil hand of Neanderthal Man from La Ferrassie, France, 60,000 years BP. Thumb is opposable but the hand is not thought to have been as dextrous as modern man's.*

The hands themselves (Fig. 41) seem to reflect in every way what we know of the Neanderthalers' ecology and habits. Like the build of the body as a whole, the hands were broad, the knuckles and joints nobbly, the muscles exceedingly well developed, especially the thenar and the interosseus muscles, and the fingertips large and spatulate. In the matter of function there is good evidence to suggest that the precision grip was powerful and well developed. According to Musgrave's estimates based on the left and right hand of the skeleton from La Ferrassie in France, the Opposability Index (see p. 70) was 64, which lies in the middle of the range of modern man (61 to 69). From these measurements there is no doubt that Neanderthal man was able to approximate his thumb and index tips, but just how much and for what he utilised the precision grip during his normal life style is quite another matter.

Neanderthal man was a hunter and his principal chore was to dismember and cut up the carcases of the animals he killed and to clean and cut up the hides. To this end he developed a highly specialised tool kit of cutters, piercers and scrapers. In addition Neanderthalers employed more general tools such as handaxes, spear heads and single-edged stone knives, but undoubtedly the Neanderthalers' main contribution to the advance of technology was the enormous variety of flake tools which the French archaeologist, François Bordes, stated to be at least sixty in number. All such tools as these are capable of being manipulated in a power grip, indeed bearing in mind the use to which they were put, they could hardly have been employed any other way.

It was not until the Upper Palaeolithic that true "precision tools" like blades and engravers were used for working bone, antler or wood; but this does not necessarily rule out the capability of precision gripping as such in Neanderthal man, it simply implies that climatically and ecologically hard-pressed as they were, they had no time to spare from the business of survival to indulge in the fripperies of their more leisured successors.

Part Two
Social and Cultural Aspects of the Hand

Chapter Five

Tool-using and Tool-making

Tool-using and Tool-making

Benjamin Franklin was not only a distinguished writer and publisher, a prolific inventor of scientific apparatus and a wise diplomat, but he was a philosopher and humanist. Perspicacious as he was, it still comes as something of a surprise that it was Benjamin Franklin who was the first to perceive an essential element of man's cultural nature: "Man", he said, "is a tool-making animal."

As Kenneth Oakley of the British Museum (Natural History) has observed, the question of drawing a boundary between pre-humans and humans had not yet become a practical issue in Franklin's time. Man may have existed in the past but that he should have been the subject of evolutionary change was unthinkable. It was not until the last quarter of the nineteenth century after *On the Origin of Species* (1859) and the *Descent of Man* (1871) had been published that man's thoughts expanded enough to encompass the revolutionary truth that Bishop Usher might have been wrong about the age of the world, and that, on the contrary, mankind was extremely ancient and must have possessed forerunners who were less human than he was. Even so the existence of a pre-human or ape-man stage was not so readily accepted until Piltdown man came along to convince a sceptical and anthropologically naive world that the missing link had been found with the unanswerable combination of the skull of a man and the jaw (suitably tinkered with) of an ape. Tools (several flint flakes that in the event turned out to be faked) did not raise the dust to any extent, so the significance of tool-making was not an issue of great moment.

It was not until the discovery of fossil sites in South Africa that the issue really came to a head. Were the australopithecines that Dart had been collecting so assiduously since 1924, when the first discovery was made at Taung, human or pre-human, man or ape?

The first intimation of stone tools in South Africa came in 1953 when pebble tools of an early Oldowan type were found in the 200-foot terrace of the Vaal River. The question was immediately raised as to who the makers were – australopithecines or some more advanced group? In 1948, teeth more man-like than *Australopithecus*, were found at Swartkrans (where the robust australopithecine was already known) and given the name of *Telanthropus*. Here, then, was an "advanced group" who were tacitly credited with being the tool-makers. *Telanthropus* was subsequently dropped as a generic name when the fossils were identified as human. Few scientists believed that *Australopithecus* was a tool-maker, principally on the basis of brain-size which was no greater on average than that of the apes; but the power of suggestion had something to do with it. Man was the tool-maker, therefore if australopithecines weren't men, they weren't tool-makers. At the present time there is still no good evidence that australopithecines made tools. Where tools are found there are no australopithecines and where there are australopithecines, there are no tools.

It is perhaps rather surprising that a bipedal hominid with at least as big a brain as an ape's, was not a more advanced tool-user; but so it seems. Australopithecines may have relied solely on improvisation, picking up naturally occurring objects when they found them to fulfil an immediate purpose; or they may have specialised in adapting tools from animal bones for such purposes as digging, pounding, sawing and chopping. The osteodontokeratic (bone, tooth and horn) culture, as it is called (see below), is no more advanced in the technology of tool use than the behaviour of chimpanzees which employ sticks as probes, and leaves as sponges and "toilet paper".

The distinction between tool-using and tool-making is of critical importance. One might say that almost any primate can *use* a tool (in the widest sense of the word) but only a human being can *make* one; the terms must therefore be used circumspectly. The difference between using and making is largely an affair of the central nervous system and involves a qualitative shift in cerebral activity from percept to concept. Abstract thinking is not a talent of non-human primates which live on a strictly "here and now" basis and for whom the past and the future have very little meaning. Some inkling of the power of planning ahead seems to be possessed by chimpanzees, whose tool-modifying activities suggest that they occupy the grey area within the spectrum of tool-using to tool-making – a stage of dawning comprehension.

Imagination is basic to tool-making. All man-made tools start off as chunks of undifferentiated material which are then shaped according to some cerebral blueprint of what is required. Every stone tool is an act of imagination and – for Aristotle – of artistry, for, as he expressed it in *De partibus animalium*, "Art . . . consists of the conception of the result to be produced before its realisation in the material."

The definition of man-the-tool-maker is a good one, partly because it forms the basis for an objective definition of man but mainly because it recognises that behaviour as well as structure can be used, even in palaeontology, for the diagnosis of *Homo*. Tool-making is a key function in a given fossil in that it supposes the presence of a number of contingent human characteristics such as the power and precision of the hands, bipedalism, and a large brain.

The living great apes, whose hands are classed as true hands, never had a chance to deploy them as man has done. When arboreal, the ape hand was preoccupied with climbing, maintaining balance, and food collecting; when on the ground the ape hands are still subordinated to the needs of locomotion (knuckle-walking).

The most widely held opinion regarding the splitting of the African apes and human stocks is that this critical event took place before the specialised "brachiating" characteristics of the ape hand had evolved. Russell Tuttle, who holds this view, believes that before becoming bipedal, ancestral hominids used their hands in a palmigrade fashion as do monkeys, and not in the knuckle-walking posture of modern African apes. He also believes that the early hominid hand had the general proportions of modern man and not those of the long-handed, long-digited and short-thumbed apes. Tuttle proposes that selection for a specifically modern type of human hand came with tool-making. I find little in this view to disagree with.

The use of tool-making as a criterion of humanity carries the further implication that its evolution was a sudden event, an inspired flash of conceptual thought, a bolt from the Pliocene blue in fact. It suggests that overnight a divine inspiration lit upon some favoured ape-man which thereby set his tribe on the way to the stars. But it doesn't work like that. Evolutionary changes are gradual, not abrupt like instant mashed potato. Tool-making grew out of tool-using over millions of years. Study of the use of tools among animals makes it clear that given the physical means and the environmental incentive many animals, including primates, will use tools. The custom is both widespread and ancient, and it was perfectly logical that the early hominids should have joined the ranks of tool-using animals, given the criteria of physical means and environmental incentive that they undoubtedly possessed. *Man was a tool-user before he was a tool-maker.* This is certain and there is no possible way in which such an apprenticeship could have been waived. Furthermore, when he had progressed beyond simple tool-using, he is likely to have passed through a bridging stage which can best be called *tool-modifying*. Tool-modifying, as its name implies, involves adapting an existing object by simple means to improve its performance. The osteo-dontokeratic (bone, tooth and horn) culture of Dart provides an excellent example of tool-modifying.

We have now introduced the three grades of a developing technology: tool-using, tool-modifying and tool-making. Let us define them:

   I   TOOL-USING: Tool-using is an act of improvisation in which a naturally occurring object is utilised for an immediate purpose, and discarded.

  II   TOOL-MODIFYING: Tool-modifying consists of adapting a naturally occurring object by simple means to improve its performance: once used it may be discarded or retained.

 III   TOOL-MAKING: Tool-making is an activity by which a naturally occurring object is transformed in a set and regular manner into an appropriate tool for a definite purpose.

Inherent in this definition of tool-making is the maintenance of continuity in successive generations by means of example and demonstration. Whether this is what is meant by culture I am quite unable to say; certainly it is one aspect of it. If culture includes imitation as this brief extract from Goodall (1968) attests: "In the wild I saw infant chimpanzees, on many occasions, not only watching adults as they worked [at termite fishing] but also picking up and using the same tools when the adults moved away" (p. 209), then chimpanzees have a culture. Goodall observed the use of leaves as "lavatory paper" by infant chimpanzees in imitation of their mothers' toilet behaviour.

Ironically, tool-making in man must have been preceded by a lengthy period of *ad hoc* tool-making and, even earlier, of *ad hoc* tool-modifying before the technique settled down as part of an established behavioural repertoire, lost and re-discovered many times over, perhaps only becoming permanent with the evolution of speech and language.

Tool-making skill depends on both peripheral and central factors. It depends on the proportions of the hand as well as the size and complexity of the cerebral cortex. S. L. Washburn of Berkeley has emphasised that increase in brain size (a somewhat crude but useful method of estimating the overall capability of the brain in terms of motor and tactile functions, skill, memory and foresight – all of which take up brain space) is more likely to have followed than preceded tool-making, so that a positive feed-back became established. A breakthrough in tool-making is followed by an increase in the size and complexity of the cerebral hemispheres which is in turn followed by further advances in tool-making.

Clearly, though, if we want to scramble back nearer the origins of primate and – specifically – human tool-use, we must recall the ancient possessions of, first, stereoscopic vision and prehensility which provide the basis for hand–eye co-ordination that is so pivotal to the development of manipulative procedures; and, second, the long record of truncal uprightness which has been a feature of primate locomotion and posture since Eocene times, millions of years before the advent of bipedalism. These possessions provided the groundwork – the pre-adaptations as they are called – for bipedalism and tool-making. Thus, it was not merely inevitable; tool-making, when it evolved after a long period of incubation, was the culmination of an ancient primate trend involving the hands, the eyes and the brain in three-way co-ordination.

Tool use in Animals

The classic example of tool-using in birds is the case of the woodpecker-finch of the Galapagos Islands which makes use of a thorn or a cactus spine held in its beak to winkle out grubs and other insects from the bark of trees. The woodpecker-finch is said to use a certain amount of discrimination in choosing a spine, rejecting those that are too short or too flexible.

Among mammals, the sea-otters of the Californian coastal waters are famous for their appealing method of feeding. Floating on their backs they employ a flat rock balanced on their chests to act as an anvil for cracking open the hard shell of certain molluscs which form part of their diet. Holding shell-fish in a two-handed grip they smash it with powerful strokes onto the rock.

There are many examples of the "anvil" technique in nature. Thrushes, blackbirds and starlings, for instance, holding an occupied snail shell in the beak repeatedly beat it on paving stones or an asphalt path until the shell is fragmented and the snail can be extracted. The technique of the lammergeier (the Bearded Vulture) differs only quantitatively. When it needs to crack open a mammalian long-bone for its contained marrow it drops the bone in a sort of bombing run on to a large rock from several hundred feet. An example of tool-using which was in no sense an innate action, as were the examples quoted above, was the saga of Jackie. His actions reflected a certain amount of

42. Capuchin using three different methods for cracking open strong fibrous husks of the Cumare *fruit (Courtesy of Drs Kosei Izawa and Akinori Mizuno).*

a

b

c

insight in problem solving. Jackie, an old and toothless capuchin monkey in the London Zoo, was inordinately fond of nuts, particularly brazils. Using a heavy marrow bone in his hands, Jackie, with great persistence, would attack the nut until eventually it succumbed. Marrow bones do not figure much in a wild capuchin's daily round so there was nothing innate about Jackie's action. At his level he was just as brilliant an inventor as Arkwright or Edison. However, a not dissimilar action has been reported for a capuchin in the wild (Fig. 42). It is important to realise that as Jane Goodall has said in her book *In the Shadow of Man*, it is only because of the close association in people's minds of tools and men that special attention has been focused upon any animal able to use an object as a tool. No special intelligence is being displayed by sea-otters, woodpecker-finches or lammergeiers. Like Annie Oakley in the song, they are simply doing what comes naturally.

It is among the apes that definite advances in tool use are seen. Yet, oddly enough, it is only chimpanzees that show a real flair. Gibbons and gorillas show no interest in "objects". Dr George Schaller, the first scientist to study gorillas at close quarters in the wild, reports that wild gorillas totally ignore objects purposefully planted in their path. Young orang-utans in captivity, however, show a keen curiosity for mechanical objects and usually end by taking them to pieces; there is little knowledge of the implemental behaviour of orang-utans in the wild, although the use of sticks as missiles, levers to open bees' nests and weapons to kill snakes have been reported. Chimpanzees are a class apart as far as tool use is concerned. Although anatomically very close, chimpanzees and gorillas are essentially and temperamentally disparate, chimpanzees being noisier, more extroverted and infinitely more inquisitive. These aspects of their temperaments are reflected in the organisation of their respective social systems. Stretching a point a bit, one might liken that of gorillas to a well-brought-up Victorian family dominated by Papa. Chimpanzees on the other hand constitute a sort of hippy commune, happy and feckless.

At least thirteen different examples of tool use in chimpanzees have been recorded showing a wide range of implemental activity. One of the most interesting is the so-called termite fishing behaviour which was first observed by Jane Goodall in 1961 among a group of animals she was studying in the Gombe Stream National Park in Tanzania. At certain seasons of the year the chimpanzees go fishing – fishing for termites that is, using a long, slender twig or vine which is poked into the heart of the nest down the flight holes through which the winged termites emerge to engage in their nuptial flights. When the chimp withdraws the twig it is well covered with worker termites which are clinging on by their mandibles. The termites are then picked off by the lips and tongue (Fig. 43).

The most interesting part of this activity is the preliminary collection of suitable twigs from areas as much as 15 yards from the termite nests. Considerable care is exerted in choosing "fishing rods" of suitable length, diameter and flexibility and when the chimpanzee settles down for a termite orgy he has a number of spare rods beside him ready for use.

The action that elevates termite fishing from mere tool-using to tool-modifying is the careful pruning by the chimp of leaves and side-

43. Chimpanzees of Gombe Stream Reserve "fishing" for termites in mound (Courtesy of Jane Van Lawick Goodall).

branches that would impede their use as probes. If we label this sort of activity a crude form of tool-making we make nonsense of the definition of man-the-tool-maker. So, in order to restore the perspective, we must either – as Louis Leakey once said – re-define tool-making or re-define man. In this book I have preferred to re-define tool-making by creating a third category of tool-modifying within the evolutionary stream of tool-using to tool-making.

As far as the status of chimpanzees as tool-users is concerned, we must recall that the chimpanzee hand is anatomically ill-equipped for tool-making. A shortness of the thumb in relation to the great length of the fingers largely accounts for this. There are other factors too that militate against an effective power and precision grip. In this sense the chimpanzee has reached its technological ceiling. Sadly, the preparation of a suitable rod for termite fishing is about as far as it can go.

Already we are falling into an anthropomorphic trap. "Poor chimpanzees," we say, "how sad to be so deprived!" What possible incentive is there for a chimpanzee to make a tool? But why should a chimp be pitied for his inability to construct a hand-axe, an engraver, bone needles, the spear and the arrow? He has no need for objects like these.

It is only the extroverted, inquisitive personality and innate intelligence of these apes that, in captivity, make them suckers for anything new. They will romp through a zoo "tea party", pile boxes and join two sticks to obtain a reward, learn sign and symbol language without tears, and generally ham it up to please humans with unfailing enthusiasm.

The Osteodontokeratic Culture

Nevertheless the superiority of chimpanzees in tool-using over any other non-human primate receives recognition by erecting a special category of tool use called tool-modifying. As tool-modifying is a grade within the continuous process of biological improvement, early hominids must inevitably have passed through just such a stage on their way to fully-fledged tool-making. Perhaps the much maligned osteodontokeratic (bone, tooth and horn) culture of Dart is just such a case; it has a lot in common in terms of the complexity of activity with the inspired ingenuities of Jane Goodall's Gombe chimpanzees.

In a memoir published in 1957, Professor Raymond Dart described the vast accumulation of animal bones, teeth and horns discovered at Makapansgat, mostly (92 per cent) bovid in origin. He argued on statistical grounds that the evidence of selection of the material (as opposed to a chance accumulation) could not be explained simply in terms of animal predators and scavengers. Instead Dart proposed that it was the work of the gracile australopithecines, themselves hunters and animal killers.

The selectivity of the accumulations, which was Dart's main argument, is exemplified by an over-representation of lower jaws, cannon bones, the distal portion of the humerus and a large number of bony flakes derived from spiral fractures of long bones thought to be deliberately produced. Other parts of the postcranial skeleton were under-represented. One criticism levelled at selectivity is that there is a striking similarity in the parts that survive in different sites where bones have, unequivocally, accumulated by natural means. Dr C. K. Brain reports on the bones of goats fed on by Hottentots and their dogs in the Namib and which were then discarded to lie on the desert surface around the Hottentot villages. Calculations of the percentage survival of each part was made and when the results were compared with the percentage survival found by Dart at Makapansgat, the similarity is striking. Brain's view is that, like the natural accumulations of goat bones, the contents of the Makapansgat accumulations are entirely predictable and dependent on the intrinsic strength characteristics of each bone. He concludes that no tool-orientated selectivity need be inferred, and with that the main prop of Dart's argument appears to fall to the ground.

The Makapansgat accumulations are found imbedded in the solidified breccia of limestone caves and the question arises as to who or what was responsible for bringing them there in the first place. If australopithecines were not the chief agency, who or what was? Porcupines are prime suspects, as they are inveterate "bone collectors"; the other suspect, the hyaena, is not a collector; it feeds out on the savannah where the food is. This leaves carnivores such as leopards or sabre-tooths which, according to Dr C. K. Brain of the Transvaal Museum, predated australopithecines. Even if leopards were not the only agency, they – at least – may have been responsible for the punctured and crushed australopithecine skulls, particularly of juveniles. It is easier to accept this interpretation than go along with Dart's belief that australopithecines were killers given to

ABSOLUTE AGE BP	GLACIAL PERIODS	AFRICAN AND EUROPEAN STONE INDUSTRIES	GEOLOGICAL SUB-EPOCHS		ABSOLUTE AGE BP
10,000	POST-GLACIAL	Iron, Bronze, Neolithic & Mesolithic industries	HOLOCENE		10,000
50,000	WURMGLACIAL	Blade-tool industries (e.g. Aurignacian)	Middle & Upper Blaeolithic	UPPER PLEISTOCENE	50,000
150,000		Mousterian Tayacian			150,000
250,000	3RD INTERGLACIAL			LATE	250,000
	RISS GLACIAL				
350,000	GREAT INTERGLACIAL				350,000
450,000	MINDEL GLACIAL		MIDDLE PLEISTOCENE		450,000
550,000	1ST INTERGLACIAL			EARLY	550,000
650,000	GUNZ GLACIAL	African only			650,000
750,000					750,000
850,000					850,000
950,000			Lower Potaeolithic		950,000
1,050,000	POSSIBLE EARLIER GLACIALS				1,050,000
1,150,000					1,150,000
1,250,000				LOWER PLEISTOCENE	1,250,000
1,350,000					1,350,000
1,450,000					1,450,000
1,550,000					1,550,000
2,000,000					2,000,000

Stone industry arrows (vertical labels): Pebble-tool industries (e.g. Oldowan); Hand-axe industries (e.g. Acheulian); Flake industries (e.g. Clactonian)

44. *A correlative scheme bringing together glacial, lithic and geological subdivisions of the Pleistocene.*

internecine murder and cannibalism.

Professor Phillip Tobias of the University of the Witwatersrand has over the years studied Dart's evidence most carefully. At one time he attributed the collections entirely to human activity (without homicidal overtones), but now he is inclined to take the view that a multiplicity of agencies were responsible – humans, hyaenas, porcupines, leopards and sabre-tooths. As Tobias has said to me, there is a strong *prima facie* case to expect that australopithecines would have at least as much implemental potentiality as the anthropoid apes, if not more. We have already seen that chimpanzees, or at least some groups anyway, are tool-modifiers. Among the osteodontokeratic tool kit exhaustively analysed by Dart there are plenty of examples of modification of "a naturally occurring object to improve its performance": horn-cores, which are used as wedges to split open the shafts of long bones; the modified humerus with its proximal extremity removed, to be used as a bludgeon – the characteristic weapon of the gracile australopithecines; and even "scoops" made from cannon bones designed to make life easier for edentulous australopithecines! There is sufficient evidence among the animal bones in the accumulations that at least some of them were the work of *Australopithecus*.

Abbé Breuil, the noted archaeologist, in a discussion of the bone accumulations at Choukoutien, near Peking, stated that "a Bone Age preceded a Stone Age". This may well be so, as the phylogeny of implemental use (from tool-using to tool-making) suggests, but there is also strong evidence from different parts of the world that bone industries continued to flourish parallel with stone industries throughout the Pleistocene (Fig. 44). For instance, the cave sites of early Pekin Man at Choukoutien showed evidence of broken bones being used as implements. Indeed the modification of bone and antler has continued to this day. Still to be picked up in antique shops are nineteenth-century "apple-corers" made from the cannon bones of sheep. These bony bygones, much used in farmhouse kitchens, are almost identical with scoops found at Makapansgat and attributed by Dart as feeding aids for toothless australopithecines.

Stone Cultures of the Plio-Pleistocene

The seeds of doubt that man was more ancient than the Last Flood began to pervade scientific thought in the 1830s when flint tools were

45. The pebble-chopper, fore-runner of the hand-axe, represents the most primitive form of fabricated stone tool. Pebble-choppers were in use for at least one and a half million years. Courtesy of the Trustees of the British Museum (Natural History).

found associated with the bones of antediluvian animals in cave deposits in Belgium, in Kent's cavern, Torquay, at Abbeville, in the ancient gravels of the Somme and, later, at St Acheul, near Amiens. In the same year as Darwin's *On the Origin of Species* was published, Joseph Prestwich, the eminent geologist, gave a paper to the Royal Society, freely accepting the evidence that man was no post-catastrophic new boy but an ancient inhabitant of the Western world. Where there were stone tools, there had been men; so the hunt for what Boucher de Perthes called "these rude stones" was on. They came to light in their thousands from all epochs, purportedly even as far back as the Eocene, the dawn era of primates! Some specimens of course were legitimate and easily recognisable tools such as Acheulian hand-axes and other well established tool-types from the Pleistocene, but the majority (including the Eocene examples which were originally called eoliths) have not received universal acclaim, being regarded (quite rightly as we now know) as accidents of nature rather than artifacts of man. Eoliths are products of natural agencies resulting from river or wave action, frost and soil creep, in which powerful friction of one stone against another produced flakes that

46. (a) *Aurignacian burin held in a precision grip.*
     (b) *Pebble-chopper held in a power grip.*

mimicked such well attested stone implements as blade tools, Mousterian scrapers and pebble-choppers.

There is no reason why eoliths should not have been used on an *ad hoc* basis by early man. Oakley comments that naturally occurring flakes are used by some Australian aborigines to this day to fashion wooden objects. It is possible too that some eoliths are indeed the work of man, but how is it possible to distinguish the artifacts from the natural objects? As a French archaeologist talking about eoliths remarked: "Man made one, God made ten thousand – God help the man who tries to see the one in the ten thousand."

The earliest undoubted standardised artifact is the pebble-tool, or pebble-chopper as it is now called, that dates back at least to the end of the Pliocene, 2·8 million years BP, at Koobi Fora, Kenya, and probably earlier still. The best known site is Olduvai Gorge where, for the first time, living sites were discovered with occupants, food bones and implements side by side. The potassium-argon dating technique puts the time of the earliest living sites as 1·9 to 1·75 million years BP. Pebble-choppers (Fig. 45) are made out of cobbles from which a number of flakes have been crudely struck off from both faces to form a cutting edge where the flake scars meet. The butt of the implement which is not worked is smooth and rounded and fits comfortably in the hand. Other tools such as spheroids and flakes were present in the same deposits but undoubtedly the pebble-

47. Author attempting to construct a hand-axe in flint by stone-on-stone method using power grip.

chopper was the principal tool and was used in a power grip (Fig. 46b).

To the untrained eye, pebble-choppers look like naturally flaked stones, but there are at least two compelling reasons why they are unquestionably artifacts. First, a similar flaking technique is repeated with remarkable consistency on thousands of specimens showing, in Leakey's words, "a set and regular pattern"; second, the material used does not always occur naturally at the site. The raw material used at Olduvai includes lava and quartzite which must have been conveyed there by human agency. Such introduced materials are called "manuports" by Mary Leakey. This type of experimental technique was developed in the first place by Dr Grover Krantz, the expert on the footprints of the Sasquatch, or "Bigfoot", of North America.

The crudest method of preparing a pebble-chopper is by the stone-on-stone technique. The hammerstone is held in one hand and the pebble in the other, or it can be rested on the knee (Fig. 47). It is not easy using this method to exert any fine control. Even the earliest Acheulian hand-axes were made by this technique.

The pebble-chopper, the characteristic implement of the Oldowan culture, persisted well into the Middle Pleistocene, but hot on its heels was its derivative – the hand-axe. Pebble tools were still widely employed when the first hand-axe was made in Africa at least 1·5 million years ago and they continued to develop side by side. It is generally agreed that the hand-axe was an extension of the pebble-chopper technique. Hand-axes to begin with were made by the stone-on-stone method and the resultant flakes were crude and deep-biting; by alternating the flaking process from one surface to the other, a sinuous or zig-zag cutting edge was achieved. As hand-axes continued to improve over the next 1·25 million years, they assumed a more oval or pearshape (Fig. 48). Flaking was delicate and shallow and, consequently, the cutting edge became less sinuous. A "hammerstone" in the shape of a wooden baton or an antler replaced stone, to be followed by the even more precise technique of pressure-flaking.

Hand-axes were a general purpose tool – a scout-knife – that could cut, scrape and stab the skins of wild animals preparatory to skinning. They were not hafted but held in the hand in a power grip. As with pebble tools, it is perfectly possible to make a crude form of hand-axe using the power grip only (see Fig. 47), thus demonstrating that the "tools are as good as the hands that make them".

Tool-using and Tool-making

The Lower Acheulian in Africa is represented in the fossil remains of *Homo erectus* at Olduvai Gorge in East Africa and Ternifine in Algeria. Outside Africa advances in stone-tool use among *Homo erectus* populations were much slower. For example no artifacts have been found in Java, and at Choukoutien, the home of Pekin man, only pebble tools and flakes of a Developed Oldowan culture are known; Acheulian hand-axes never appear. In Europe the earliest recorded hand-axes are from Nice in the south of France dated 400,000 years ago. The hand-axe influence spread slowly northward reaching England at Swanscombe, Kent, and Furze Platt in Berkshire by 250,000 years ago. The hand-axe culture persisted in Europe until the first warm interval of the last glacial period.

Flakes have probably always been used whenever any form of stone knapping was practised. It would be natural to make use of the debris of sharp-edged slivers of flint, chert or obsidian even if the construction of a core-tool was the primary consideration. Flint flakes can be used for all sorts of modern purposes, including shaving and preparing a hare for the pot, as I have proved to my satisfaction. Thin sharp slivers of flint which could be inserted into the joints to sever the intra-articular ligaments were particularly valuable. Even at the site of primitive pebble-chopper industries, some of the flakes provided evidence of use.

It is a matter of surprise to me that a true "flake industry" did not evolve earlier than it did. The Clactonian (circa 400,000 years ago) was the cultural period when craftsmen started to specialise in flake tools, converting them into scrapers, knives and even "spoke-shaves" for shaping wooden spears (Oakley, 1967). The Clactonian people appear to have reversed the hand-axe tradition by keeping the flakes and throwing away the cores, as at Swanscombe's lowest levels where hand-axes are absent. In the upper gravels of the River Thames at Swanscombe, hand-axes return to the scene as people steeped in the Acheulian tradition supplanted the Clactonian tool-makers. The Clactonian exerted a powerful influence, as it seems very probable that it developed into the Mousterian culture which kept Neanderthal man alive for 70,000 years (see Fig. 44).

The Upper Palaeolithic industries (Chatelperonian, Aurignacian, Gravettian and Magdalenian in order of succession) collectively known as *Blade tool cultures* represented a vast and very sophisticated leap forward in the social and technological history of man. *Homo sapiens* was now on the scene. He had evolved a large brain, small

48. Hand-axes showing refinement and improved technique between Lower Acheulian, 500,000 years ago (top) and Upper Acheulian, 100,000 years ago.

South of
WADY-SIOR
Palestine

Tool-using and Tool-making

teeth, and dextrous hands morphologically identical with those of modern man. The best known of these populations were the Cro-Magnons who had spread from Southwest Asia into Europe where geographically and culturally they replaced the Neanderthalers. Cro-Magnons exhibited, as Oakley says, "... a remarkable aesthetic sense and displayed artistic skills scarcely excelled in any later period". Their use of materials had expanded to include bone, antler and ivory as well as wood and flint.

The characteristic implements of the Upper Palaeolithic, a far cry from the relatively crude chopper–hand-axe tradition, consisted of delicately wrought blades of stone, which included knives, chisels and engraving implements called burins. The Cro-Magnons also made harpoons of antler, and needles of bone, "Venus" figures (Gravettian) shaped from limestone, and, above all they were responsible for those artistic wonders of the ancient world – the Franco-Cantabrian cave-paintings.

The nature of tool-making was a very simple one up to the end of the Mousterian – a matter of choppers, axes, and flakes which had preoccupied men of the Old Stone Age for nearly 2·25 million years – an incredibly long period of technological stagnation. However, the progress of stone technology must be judged against the background of the life-style of the tool-makers themselves.

The most important aspect of the life of Old Stone Age man was food-getting – hunting or scavenging of meat and the gathering of berries and edible roots. Primitive man was preoccupied with this need from dawn to dusk and the implements he used reflect his single-mindedness. There was no time in his daily round for such trivia as clothes, utensils or artistic embellishments to cave walls. These activities only came about when populations were larger, when hunting techniques were more efficient and less time-wasting, and when the discovery of fire served, among its many functions, to prolong the length of the working day. Spare-time was a luxury that the Olduvai people, the *erectus* populations and even the early *sapiens* group could not afford. It was not until the last glacial was virtually over and game was plentiful once more that time could be spared for the more leisurely, less vital activities.

The gorilla, unlike the chimpanzee, does not use tools; he has plenty of spare time, but lacks incentive. Early man, on the other hand, had plenty of incentive; what he lacked was time.

Chapter Six
Handedness

Handedness

Tibetans have a saying: "Beware of the devils on your left hand side", a maxim common to every race and creed in one form or another. In most cultural groups the left hand is regarded as both unclean and unworthy. Among Middle Eastern and North African countries the ritual of eating with the right hand only is strictly adhered to in orthodox circles; an awkward procedure at the best of times, but taken all in all it is probably an important hygienic one, since it is the left hand that is often used exclusively for toilet purposes. In spite of strict injunctions about washing, the left hand must always be defiled, philosophically if not bacteriologically, by the excremental tasks it is called upon to perform.

In the English language the very words "left" and "right" carry their own commendation or detraction as the case may be. The right implies correctness and propriety, the left in its primary sense, means weak and worthless. Many derived words like righteous and dextrous, gauche and sinister retain their original etymological significance – skill and goodness versus incompetence and evil. Politics, of course, are a special case, although there are those who might not admit to the distinction. According to H. L. Mencken, the use of the Left and Right in politics stems from the seating of the members of the French Assembly in 1789; the conservatives of noble birth sat on the right of the presiding official, the revolutionary members of the Third Estate to his left, and the moderates in the middle.

In the north country dialects, cack-handed means left-handed. To cack, a word in common use in medieval English, means to defecate. The implication of cack-handed therefore is that the individual, left-handed, employs the hand used dominantly in toilet procedures. One wonders if those in whose vocabulary cack-handed still holds a place, realise that they are using the medieval equivalent of a modern four-letter word? According to Partridge's dictionary, cack has a number of variants in different areas of Britain: cork-handed appears in Derbyshire, gallock-handed in East Yorkshire; bang-handed in the Tyne and Tees area; keggy in Castle Bromwich; wacky in Evesham; spuddy-handed in Gloucestershire; scrammy in Bristol; kefty in Somerset; corrie-pawed or corrie-fisted in different parts of Scotland.

From the dialectal obscurity of Devonshire, Norman Capener, scholar and orthopaedic surgeon mentioned more than once in this book, resurrected "couchie" which he suggested was a corruption of *gauche*. Barsley in his excellent little book – *The Left-handed Book* –

has collected some of the words commonly used in foreign languages showing that the attitude towards left-handedness and left-handers is no less derogatory and disapproving than here in Britain.

Handedness, like bipedalism and tool-making, is popularly supposed to be a unique possession of man. However, although it may be true that most animals (with notable exceptions, the male fiddler-crab with its giant right pincer for example) do not overtly demonstrate any lateral preference, there seems to be no doubt that higher primates, at least, show some degree of handedness.

Ambidexterity means the equally skilled use of both hands, but it should be implicit in the definition that the hands are skilled in the first place. The term therefore should probably not be used except where some degree of skilled handedness is established. A horse can hardly be said to be ambidextrous although one might recall, somewhat irrelevantly, the famous case of Clever Hans, the German Wonder-Horse, which astonished audiences by accurately tapping out a number selected by the audience using either front hoof.

From horses to the drawing of horses, one thinks naturally of Sir Edwin Landseer, one of many artists reputed to be ambidextrous; in his case the evidence that he possessed this talent is unequivocal. History relates that he could draw a portrait of a stag with one hand while simultaneously drawing the likeness of a horse with the other, neither being the less perfect. The ambidextrousness of Leonardo da Vinci is a matter of perennial argument. Certainly Merekowski, his biographer, does not believe him to have been ambidextrous. Norman Capener, an expert on the anatomy and surgery of the hand, has expressed some doubt as to whether Leonardo possessed any inborn left-hand skills at all, over and above those normally possessed by man. It is known that a brush with a gang of muggers left his right hand badly mutilated. He may certainly have trained his left hand to take over some of the duties of draftsmanship from his mangled right. There is no doubt that, short of full-blooded ambidexterity (of the Landseer pattern), which is rather rare, individuals possess different degrees of ambidextrousness and in such cases transfer of tasks from one hand to the other would be a relatively easy progression.

Capener is in the minority when he asserts that Leonardo was right-handed; most authorities support the sinistral theory, including Merekowski who believes that Leonardo drew and wrote with his left hand and painted with his right. Leonardo's mirror-writing in his *Notebooks* is often quoted in support of left-handedness but may be quite irrelevant. He may have used mirror-writing not because he was left-handed but because he wished to keep his private thoughts, some of which were said to be extremely radical, to himself. Barsley suggests that Leonardo was a wayward genius who liked to puzzle others – another possible explanation. Other evidence of left-handedness includes the positioning of the cranks in his mechanical drawings such that they could only be used by the left hand. This evidence, perhaps, is more telling but unfortunately does not indicate which hand he *painted* with. Mixed-handed individuals (as distinct from those truly ambidextrous) may be right-handed writers but left-handed manipulators. The problem of Leonardo's handedness remains an unsolved mystery.

It has been estimated that there are approximately 200 million left-handed people throughout the world. This probably represents about 5 per cent of a population of 4,000 million. Such a figure, however, is fairly meaningless for a number of reasons, one of which is that it does not distinguish between the frequency of left-handedness at birth and in adult life; in other words there is no means of knowing whether or not the estimate takes into account the infant left-hander who has been forced for various reasons into changing his ways. Dr Bryng Bryngelson of the University of Minnesota estimates that if there was no interference by parents and teachers "34 out of every 100 children born today would become left-handed". Left-handedness – or rather the toleration of left-handedness – appears to be on the increase. A few years ago 4 per cent was considered reasonable within a given sample; nowadays the figure has risen to between 5 and 10 per cent. There are obviously many factors that would influence the frequency apart from parental and teacher influence, such as intelligence, mental retardation, race and even national ideology.

It may seem strange to include national ideology as a formative factor in handedness, but consider for a moment the likely attitude of a totalitarian regime toward the mavericks of the state who persist in resisting the precepts of national conformity. One can imagine the dictat to all schools that pupils be taught to employ their *right* hands. Non-conformity among the young, it would be said, provides the breeding ground of dissidence. Without the appropriate statistics one would suppose that the Iron Curtain countries show the lowest rate for left-handedness (except possibly the People's Republic of China). In the United States, Britain, most Western European countries (and

Israel, according to Barsley) considerable tolerance is shown. Perhaps this is true at an overall national level but it would be interesting to know whether there is a similar tolerance, in Britain for instance, in conservative and elitest institutions. Are there any left-handed Wykehamists? Does Sandhurst permit the southpaw? Or crossing the Atlantic, would the FBI tolerate the sinistral-handed among its agents.

One of the largest series of statistics available is that of Karpinos and Grossman (1953) who analysed the manual preferences of 12,159 male US Army recruits, including rejects as well as successful candidates, with the following results:

R. Hand preferences	91·4%
L. Hand preferences	8·0%

The investigation was carried out by questionnaire, generally considered to be the fairest method. A significant fact emerged and that was that left-handedness was more frequent among the rejects than among the acceptances. A low IQ has often been regarded as consonant with left-handedness but a recent study by Newcombe and her colleagues refuted this. Indeed the IQ scores in Newcombe's series tended to be lower among those classed as pure right-handed than among the mixed-handed group. On the other hand, Barsley states that the incidence of left-handedness in mental institutions may be as high as 30 per cent.

Handedness in Non-human Primates

The first statistical enquiry into handedness in apes was probably Finch's study of thirty chimpanzees carried out in 1941. The animals were required to obtain a food reward by means of manipulatory tasks. Each animal was given 800 trials. The results at 80 per cent and 90 per cent levels were as follows:

	In over 90% of trials	*In over 80% of trials*
R. handedness	30%	36·6%
Ambiguous	40%	16·8%
L. handedness	30%	46·6%

Handedness

The figures reveal, first, that chimpanzees show well-marked handedness (60 per cent in the first column above and 82 per cent in the second column, discounting "ambiguous" results). Lateral preference is not clear-cut but if anything there is a slight preference for the left.

Helmut Albrecht and Sinclair Dunnett's West African chimpanzee series was even smaller, but the point of special interest is that it provides the only figures available for wild chimpanzees. Handedness, as in Finch's study, was clear-cut in ten out of thirteen animals. Lateral preference however was equivocal, four being right-handed and six left-handed. The task that Albrecht and Dunnett observed was grapefruit peeling; the fruit was held in one hand, and the contralateral thumb was used to strip off the skin of the fruit.

Most of the experimental work on monkeys has been carried out in macaques (*Macaca mulatta* – the rhesus, and *M. fuscata* – the Japanese macaque). Warren in 1953 performed tests for handedness in eighty-four rhesus monkeys. The results of Warren's study were as follows:

	In over 90% of trials	*In over 80% of trials*
R. handedness	27%	31%
Ambiguous	46%	35%
L. handedness	27%	34%

These figures show handedness is present in macaque monkeys but less strongly so than in chimpanzees (see above). For example at the 80 per cent level in chimpanzees, handedness is present in 82 per cent while at the same level in macaques the figure is 65 per cent. A weak left-handed preference is shown which is probably not significant.

A number of recent studies have broadly confirmed Warren's findings but, with Japanese macaques at any rate, there is a significant preference for the left hand. Tokuda's investigation of forty-one free-ranging monkeys gave the following distribution of preference:

R. handedness	20%
L. handedness	39%

Comparison between the results obtained by different investigators is complicated by the nature of the tasks used to determine lateral preference since the trend towards left-handedness diminishes as the tasks increase in difficulty. The experiments with free-ranging Japanese monkeys suggest considerably stronger left-hand preferences to those macaque monkeys tested in controlled laboratory conditions.

The Origins of Right-handedness in Man

Modern man, when all the cultural forces surrounding him are taken into account, is indisputably right-handed. Such monkeys and apes as have been studied show a left-hand bias. The bias is stronger in the case of macaques than chimpanzees. Clearly, therefore, there has been a major revolution in handedness during the course of primate evolution from monkeys to apes, and apes to man. The evidence for lateral preference in early man is pretty slim and stems largely from studies of the characteristics of his tools and the hints supplied by cave paintings; thus guesswork is rife. As Thomas Carlyle wrote: "Why [the right hand] was chosen is a question not to be settled, not worth asking except as a kind of riddle." Speculation is the name of the game.

Apart from rather doubtful evidence that a number of fossil baboon skulls from South Africa, contemporary with *Australopithecus*, show signs of a blow from a right-handed adversary (early man?), most of the evidence of handedness is derived from large collections of human stone artifacts. Oakley reports that the majority of tools from Choukoutien, the site of discovery of the representatives of *Homo erectus*, were chipped by right-handed persons. Sarasin (1918) studied Mousterian scrapers and points and found that there was near-equality in lateral preference:

$$
\begin{array}{ll}
\text{R. handedness} & 34\% \\
\text{L. handedness} & 32\% \\
\end{array}
$$

However there are a number of flaws in his work which become apparent when the function of the tools is taken into account; Sarasin based right- and left-handedness on the manner of grasping alone. Posnansky was more concerned with the functional aspect – the use to which the tools were put. His work was done on a collection of 118 hand-axes from Furze Platt near Maidenhead (now housed at Wollaton Park Museum, Nottingham). By assessing the symmetry or asymmetry of the hand-axe, particularly the positioning of the median ridge marking the thickest part of the axe, Posnansky was able to sort the left-hand from the right-hand axes. He also isolated a third group where the hand-axe was more or less symmetrical and could have been used by either hand. When tabulated, Posnansky's results are as follows:

R. handedness	31·3%
Used by either hand	44·9%
L. handedness	17·0%
Equivocal	6·8%

Thus, this study points strongly to a right-handed bias in the Lower Palaeolithic at least 200,000 years ago, but there was still a long way to go before it becomes an overwhelming preference. The emphasis on right-handedness is evident when specialisation in such tools as the sickles and the scythes made it desirable that – for pragmatic reasons – there should be a one-handed orientation of society. Agricultural tools were the possessions of the community, not of the individual, and evolutionary pressures would have operated strongly on the genetic basis of handedness in keeping with the community norm. Be right-handed or die of starvation.

The final surge that turned a right-handed bias into the right-handed dominance of modern man is quite unknown. Possibly it was the pressure of community tools, perhaps the moral force of religio-mystical practices or the influence of "military training" which taught the men the need to protect the left side of the body, where the heart lies, in hand-to-hand combat with penetrating weapons. Perhaps it was the women who found it more advantageous when nursing a baby in the crook of the left arm to continue their chores with a free right hand; according to Desmond Morris, 80 per cent of all mothers cradle the baby on the left-hand side. This figure is confirmed by a recent survey of paintings of the Madonna which estimates that in 83 per cent of the paintings the left arm is doing the cradling. This pattern occurs equally frequently among left-handed mothers. Thus it is possible that purely practical considerations, from the mother's point of view, are not as important as I have suggested above. The regular beat of the heart (heard on the left) has been proposed as the important factor. The human heartbeat is already a familiar sound to the infant at birth.

Depictions of the human hand in cave art have been used to "side" the dominant hand among Upper Palaeolithic populations (*circa* 22–28 thousand years ago). According to the Abbé Breuil, hands constitute the earliest examples of cave art at Gargas and El Castillo and other sites in the Franco–Cantabrian province. At Baume-Latrone there are pictures of hands, mainly left, printed with the finger dipped in red clay. At Gargas the technique is different: the depictions are silhouettes outlined in black manganese oxide or in red ochre (iron oxide); these are "negatives". A hand dipped in paint and used as a stamp is a "positive". Negatives are usually of the left hand and positives of the right. At El Castillo there are forty-five negative hands of which thirty-five are left and nine right. If one assumes that the dominant hand was manipulating the red ochre there was a dextral dominance of 5 : 1. Contrary findings of the frequency of right profiles of animals over left profiles seem to suggest to Walter Sorell a left-hand bias. He is assuming that the cave-painters had no model and were drawing from memory, which is probably correct as far as it goes, but it is a little dangerous to assume that left-handers always draw right profiles and vice versa.

A unique feature of the stencils of hands at the Gargas cave is that in many cases the terminal bones of fingers are missing. The custom of cutting off parts of the fingers as a sacrifice or a sign of mourning is prevalent in many primitive societies. Since these mutilated hands are always left hands, man must have been right-handed as he would hardly mutilate the hand that he depended on for his daily existence. It is impossible to draw any scientifically valid conclusions about hand preference from the sort of circumstantial evidence that is derived from cave-paintings and stone tools.

The most that can be said about the origin of right-hand dominance in man is that the left-hand bias of non-human primates appears to have been converted into a right-hand dominance during the evolution of man; and that this shift was likely to have been a very slow process which did not manifest itself fully until the life-style of early man started to include activities which required a high degree of delicacy and precision in their performance. A similar sort of argument has been adduced in favour of the gradual improvement (particularly in terms of the precision grip) of the hand itself during the Pleistocene. Thus it may eventually become apparent that *handedness* and *handiness* are interdependent. Without full *handiness*, a dominant *handedness* is really irrelevant.

Chapter Seven
Fingerprints

Fingerprints

Fingerprints are the positive impressions of the papillary ridges. The occurrence, development and function of papillary ridges has already been discussed; here we are concerned with the application of fingerprint techniques in the forensic, archaeological and medical sciences. The science of fingerprint analysis is known as dermato-glyphics, a term introduced by Professor Harold Cummins of the USA in 1926. The more archaic term for the science, dactylos-copy, is still sometimes used.

Scientific specialists who study fingerprints are extraordinarily diverse in their interests. Geneticists study prints as heritable structural traits or characteristics that can be recorded. Physicians study them for their diagnostic value in disease such as Coeliac Disease and certain congenital hormonal conditions such as Mongolism, Turner's Syndrome, and Klinefelter's Syndrome. Archaeologists note fingerprints diligently because of their occur-rence on ancient clay pots and on the fingers of Egyptian mummies. Finally there are the forensic scientists and fingerprint officers whose concern is the identification and classification of accidental "dabs" found at the scene of the crime.

The first thing to understand is that fingerprints are immutable and persistent. They can be obliterated temporarily but eventually their original pattern is restored. They are also wholly individual. No instance has yet ever been noted where two separate digits possess the same papillary ridge detail.

Papillary ridges, once established, do not change with age and they are totally unaffected by environment although they undergo atrophy with loss of pattern in Coeliac Disease and severe eczema, and in skin deprived (for one reason or another) of its nerve supply they almost completely disappear.

Ridges appear in the foetus between the third and fourth months and it is at this time that the influence of congenital defects of the chromosomes on papillary growth patterns manifests itself. Palmar prints – as opposed to fingerprints – are particularly sensitive to such influences, particularly at the site of the atavistic palmar pads at the base of the fingers.

Identical twins are derived from the division of a single fertilised ovum (monozygotic). Non-identical twins (dizygotic), who have the same chance of looking like one another as any brother and sister, are born from separate ova, each fertilised by a separate spermatozoon. It is a well-established belief that identical twins are identical down to the fingertips, but the fact is that although superficially there is a tendency to overall similarity in pattern, their papillary ridge details are quite distinct (Fig. 49). The natural law that no two fingerprints are exactly alike holds for identical twins as for others.

Although the acknowledged inventor of the present-day system of classification and identification was E. R. Henry (later Sir Edward Henry, Commissioner of Police at Scotland Yard), he was not the first to be aware of the forensic function of fingerprints. Identification by means of fingerprints presupposes the existence of a fingerprint collection for comparative purposes; this is the hard meaning of identification. In the soft sense, identification of fingerprints can also be regarded in the same light as a seal impressed on sealing-wax – recognisable as the trademark of an individual. The earliest examples of fingerprints used as seals date from the third century BC when they were quite commonly used by Chinese businessmen.

Prints that have been left accidentally in clay before firing or have been used to embellish the pottery with a design such as a pastry-cook

49. Police record of a set of fingerprints of identical twins (Courtesy of the Editors of Fingerprint Whorld*).*

might employ to crimp the pastry, are referable to even earlier dates. Nail-marks were also used by potters as ornamentation and they are to be seen on the rough surface of Assyrian bricks.

In more recent times there are numerous examples of finger or thumb impressions used for authenticating documents. Thomas Bewick (1753–1828), the well-known engraver and miniaturist of natural history subjects, "signed" his works with a fingerprint and the words "Thomas Bewick, his mark".

After the Orange War the citizens of Derry petitioned King William in London for damages suffered during the siege; the document was authenticated by the thumb and fingerprints of the local worthies and (as if to prove that they were well educated and respectable men in spite of the crudeness of their marks) they also appended their signatures.

The awareness of pattern in fingerprint seems to have first wakened in the breast of Malpighi (1628–1694) whom anatomists will know from his "corpuscles" and his "layer". There is the story of an apocryphal food-chemist who in America invented a series of soft drinks called 1–Up, 2–Up and so on, all of which failed. When 6–Up failed too, he gave up his quest on the brink of success. Malpighi also missed immortality by a hair's breadth, and instead it was Purkinje (1787–1869) of the University of Breslau who is credited with the first classification of fingerprint patterns.

However the true inventors of the modern police method of classification are Sir William Herschel, Sir Francis Galton and – the only policeman among them – Sir Edward Henry. The popular system of identification in police work at the end of the nineteenth century was Bertillon's "Anthropometric System", which included *inter alia* the recording of fingerprints, but in 1901 the clumsy and subjective elements of Bertillon's method were dropped in the United Kingdom in favour of a system based entirely on fingerprints. In 1901 the first Finger Print Bureau at Scotland Yard was inaugurated. The first conviction in Britain employing fingerprint evidence was in 1902. Since then there has been a steady advance showing the growing power of this new weapon in crime prevention: the first case of murder that rested on fingerprint evidence (1905), the development of a telegraphic code for fingerprint formulae (1921), the single fingerprint system (1930), the use of palm prints (1931). The first murder conviction on the basis of palm prints took place in 1942, the first fingerprinting of a whole town (Blackburn) in a murder case in 1948 in which the culprit was eventually caught and hanged.

Acceptance by law of the evidence of fingerprints came in 1948 when the Criminal Justices Act was passed. Section 39 accepted that fingerprints "shall be sufficient proof of a person's previous conviction".

One of the unsung pioneers of dactylography was the Scottish doctor, Henry Faulds (1843–1930). Quite independently, while a medical missionary in Japan, Faulds came to the conclusion that the ridges and patterns showed a high degree of variability and were themselves immutable. In this he foreshadowed the work of Herschel, although it was Herschel who subsequently received much of the credit.

On 28 October 1880 Faulds wrote a letter to *Nature* which was (with the exception of Dr Nehemiah Grew's treatise in 1684) the first paper ever published on fingerprints. In it he recounts how his fingerprinting method actually succeeded in solving a crime. An incident took place near his home in Tokyo during which a thief, climbing over a whitewashed wall, left a set of sooty fingerprints behind. A suspect was arrested. Faulds took his fingerprints and compared them with those at the scene of the crime. They did not match and the suspect was released, thus showing how fingerprints may protect the innocent as well as bring the "villain" to justice. Later the real offender was apprehended and this time Faulds was able to demonstrate a perfect match.

In spite of this and other successful police cases, Faulds did not have much success with the British police when he came back to England. In fact they regarded him as a bit of a crank and preferred to adopt Bertillon's system rather than the ten-finger method which Faulds offered to set up for them at his own expense. Whatever Faulds may have thought of policemen, he would no doubt have liked to have known that J. Edgar Hoover, the prestigious Director of the FBI, was an ardent admirer of his work and methods.

The anatomy of the papillary ridges is determined partly by "nature" the heritable component – and partly by "nurture" – the environment. Simple pressure is clearly not conducive to the production of dermatoglyphics. The human heel pad which bears the greatest weight, never shows papillary ridges, nor do callouses on the hands of heavy manual workers. However, the baboons and others like macaques and mangabeys who walk in a *digitigrade* fashion bearing the weight on the palmar pads at the base of the fingers, show well developed whorls (Fig. 18), so the matter is not quite so simple as it may appear at first sight.

Ridge configurations occur not only on the fingertips but elsewhere in the hand, notably the sites of the palmar pads of mammals when a "delta" or triradius (the meeting point of three systems of ridges) normally occurs. On the thenar eminence ridges tend to be longitudinally aligned; on the hypothenar eminence they are transverse or oblique.

50. The four main fingerprint types:

(a) Arch,   *(b) Tented arch,*

(c) Loop,   *(d) Whorl.*

Adapted from Dr Sarah Holt in Genetics of Dermal Ridges *(Courtesy of Charles C. Thomas).*

Fingertip Patterns

Fingertip prints are of three kinds according to the Galton classification: arches, loops, and whorls. Henry added a fourth – tents.

ARCH: Ridges run from side to side in a symmetrical arch-like fashion. True triradii are absent. Arches are rare and constitute approximately 0·5 per cent of the population (Fig. 50a).

TENT: This pattern, a variant of the former, produces a vertical upthrust with ridges arching symmetrically over it (Fig. 50b).

LOOP: Loops are of two main types: those that are directed towards the ulnar side and those directed towards the radial; the delta lies on the opposite side to the direction of the loop. Loops are formed of ridges arching sharply round a core (Fig. 50c); in some loops the arching ridges pass so close together that they give the appearance of a *whorl*. Loops are the most common configuration, constituting 63·5 per cent in males and 65·6 per cent in females. Of these ulnar loops are far more frequent than radial loops. Loops are seen not only on the fingertips but also on the sites of the ancient interdigital palmar pads, and on the thenar and hypothenar pads.

WHORL: This pattern has a well formed circular or elliptical core with a delta on both sides. One variation of the whorl pattern is the *central pocket* which is essentially a loop pattern in which a few ridges around the core are closed (Fig. 50d). Whorls are second only to loops in frequency of occurrence – 26·1 per cent in males and 23·9 per cent in females.

Ridge Characteristics

Ridge characteristics provide the basis for fingerprint identification. Ridges do not run smooth and uninterrupted courses. Some end abruptly, others bifurcate, some split to form island-like enclosures, or fork to join up with adjacent lines (Fig. 51). There seems to be no particular pattern to the occurrence of ridge characteristics except that they are most commonly seen in the region of deltas (triradii) and cores. Seven different types of ridge characteristics are recognised. Sixteen points of similarity of ridge characteristics between two prints are required by the law in Great Britain to uphold claims of identity (Fig. 52). The number of points of similarity required by law outside Great Britain varies from six in India to seventeen in France.

For reasons not at all understood, the patterns of the papillary ridges on the fingers and palm duplicate the sort of patterns seen in nature in a variety of different situations. For example the formation of desert sand dune systems many miles long seen from the air, or the sand ripples left by the receding tide on a shallow beach, are almost identical, even to the presence of all seven of the recognised ridge characteristics. Other examples include such varied phenomena as the bark of chestnut trees, the backs of male cuttlefish and the coats of zebras. The "papillary" pattern of the zebra skin bears an even closer analogy with fingerprints which are sometimes seen to bear *subsidiary ridges*. These are extremely fine ridges between two normal ridge prints; being countersunk, they leave an impression only when a strong degree of pressure is used in recording the print on paper. One subspecies of Burchell's Zebra (*Equus burchelli antiquorum*), a southern race, is characterised by "ghost" stripes – narrow, faintly outlined stripes lying between two developed ones.

This duplication in nature of the characteristics of papillary lines is

RIDGE TERMINATION	
BIFURCATION	
INDEPENDENT RIDGE	
DOT OR ISLAND	
LAKE	
SPUR	
CROSSOVER	

51. Variations in papillary ridge systems.

the original observation of two fingerprint experts of the Hertford-shire Constabulary, John Berry and Martin Leadbetter. It is difficult to believe that such a remarkable series of coincidences has occurred haphazardly; rather, one is tempted to look for some fundamental pattern of growth or form.

Ridge Counts

Since the fingerprint types may differ on each finger of individuals and, furthermore, since there are a complete series of intermediate forms separating the four main pattern types, it is impossible to classify individuals according to type. To overcome such difficulties, recourse is made to "ridge counting", using the method of Bonnevia (1924). The count is based on the number of ridges which are transected by a line drawn between the nearest delta or deltas (triradii) and the centre, or core of the pattern. Arches, lacking a delta, have a score of zero. Loops have a single delta and whorls have two, in which case the higher count is used. Ridge counts on all ten digital tips are then added together. Neither terminals – the delta or the core – are included in the count, which may vary from one to sixty. There are rules of procedure as to what constitutes the measuring point of the delta, but it is sufficient here to illustrate a simple example. Subsidiary ridges (see above) are not counted.

The Henry System and its modifications are clearly explained in an HMSO monograph. The author is the famous Chief Superintendent Cherrill whose name was an honoured one in law-abiding homes and a dirty one in every thieves' kitchen.

The position regarding racial differences is today precisely as Francis Galton recognised it in 1892 when he compared the fingerprints of the English, Welsh, Basques, Jews and Negroes. Differences between races are purely statistical and depend mainly on the numbers of loops and whorls in the populations under study. Racial dermatoglyphics is not, and cannot be, concerned with individuals. Such newspaper headlines as "Fingerprints show Nationality" are rubbish. In a recent analysis involving fingerprint classifications of over half a million individuals of all races, the findings indicate that the frequency of whorls is significantly less in Caucasians than in non-Caucasians; the largest number occurs in Chinese and Japanese. The reverse – as might be expected – is true of loops; Caucasians have the greatest percentage and Mongoloids the least.

Fingerprints

According to one authority, the retention or elimination of the palm patterns of the thenar and hypothenar areas are better indicators for the comparison of races. The only racial groups displaying more thenar than hypothenar patterns are the Negro, the Maya, the Aymara, and the Kechwa. In all the other races the hypothenar patterns are considerably in excess of the thenar.

Mongolism, or Down's Syndrome, is characterised by the presence

52. Characters of papillary ridges are used to compare fingerprints using a special formula. In Great Britain the law requires 16 points of similarity to uphold identity.

PHOTOGRAPHIC ENLARGEMENT OF THE MARK FOUND ON THE KNIFE

1 Ridge Ending (^{3})
2 Ridge Ending (^{3})
3 Ridge Enging (^{0})
4 Bifurcation (^{3})
5 Bifurcation (^{0})
6 Bifurcation (^{1})
7 Ridge Ending (^{0})
8 Ridge Ending (^{1})
9 Independent Ridge (^{1})

of an extra pair of chromosomes. The effect of this genetic abnormality is a retardation of growth in most parts of the body giving rise to multiple abnormalities including mental deficiency. Abnormalities are found in the digital and palmar lines of the hand which, itself, is short with stubby digits and incurved little fingers. Skin abnormalities include frequently the presence of a "simian crease" (see p. 48) and the absence of the distal flexure crease of the little finger.

PHOTOGRAPHIC ENLARGEMENT OF LEFT FORE FINGER
IMPRESSION ON THE FINGERPRINT FORM OF THE SUSPECT

10 Ridge Ending (^{0})
11 Ridge Ending (^{1})
12 Bifurcation (^{0})
13 Bifurcation (^{2})
14 Ridge Ending (^{4})
15 Ridge Ending (^{0})
16 Ridge Ending (^{1})

Fingerprints

The presence of a "simian crease" has been held to be indicative of congenital heart disease or rubella in the mother during pregnancy.

Aberrations of the sex chromosomes are associated with conditions such as Turner's Syndrome in which ovarian tissue is congenitally absent and Klinefelter's Syndrome in which *inter alia* the mammary glands are excessively developed in the male. Both these diseases are associated with dermatoglyphic abnormalities.

Atrophy and actual loss of fingerprints occur in a variety of local conditions – peripheral nerve injuries, eczema and old age, to mention a few. Recently a relationship between coeliac disease, a metabolic condition, and severe atrophy of papillary lines has been demonstrated.

In a fairly recent textbook of comparative anatomy the authors summed up the functional role of papillary lines in an unequivocal and rather sour paragraph: "[Papillary ridges] are useless organs which serve to convict man of animal ancestry as they have on occasion served to convict them of crime". In order to spare these misguided authors more of the remorse which surely must have haunted them ever since, I refrain from mentioning their names. Bearing in mind the role of papillary ridges in manipulation and tactile discrimination, it seems hardly appropriate to call them "useless organs", to say nothing of their genetic, diagnostic and forensic applications.

As Mark Twain, ahead of his time as he often was, correctly observed in *Life on the Mississippi* and, later, in *Pudd'nhead Wilson*, fingerprints are the only indelible signatures of man. Sorell writes in his interesting book, *The Story of the Human Hand* (1967): "We can fake and forge the written name or . . . we may change the way we sign it. Some people even change their name. Nature permits such freedom of choice, but our heredity rules the signs on our fingertips with authoritarian power for life."

Chapter Eight
Gestures

Gestures

Communication between humans is a complicated business in which all the senses play their part, vision and hearing being dominant. The chief organs of human communications are the eyes, the ears and the vocal apparatus, a combination that is unique to man. This section deals with non-verbal communication.

Gesture forms a notable part of our day-to-day lives. According to Sir Richard Paget (1869–1955), man can make 700,000 different gestures which is said to be 150,000 more than there are words listed in the largest English dictionaries. While the claim is probably extravagant, there is no doubt that sign language is a very important aspect of human communication.

Like so many aspects of anatomy and physiology in man, the use of gestures, their role and the conditions under which they are used, have been taken very much for granted. As a prime function of the human hand, the use of gesture demands more than a passing comment. The following list is not claimed as definitive but it comprises the principal contexts in which gestures are employed by man. Gestures include facial expressions and bodily movements:

—As accompaniment to normal speech.
—As a substitute for a foreign language.
—As a substitute for normal speech.
—As a substitute where normal speech becomes inaudible, disadvantageous or dangerous.
—As an accompaniment to certain professional activities, e.g. by actors, dancers and political speakers to supplement or replace the spoken word.

In moderation, gesture enhances the effectiveness of speech, indeed speech without gesture is sterile and tedious. When lack of gesture extends from the hands to the face with a suppression of facial mobility, the general "deadpan" expression is irksome. On the other hand, arms that are waved like the sails of a demented windmill are nothing but an irritating distraction. I remember some years ago a TV series of military talks by General Sir Brian Horrocks, a fluent speaker, that were notable for Horrocks' impressive battery of gestures, enthusiastic yet controlled, that were a graphic accompani-

53. General Sir Brian Horrocks – a great communicator (Courtesy of the BBC).

ment to the dramatic tales of a highly mobile war that he was telling (Fig. 53). "Talking heads", in TV jargon, come in all shapes and sizes, and the best are those whose hands and faces are alive and expressing emotions as well as facts. Cold reason and impassive objectivity may be all right for historians but are not the stuff to make compulsive viewing.

While fluent speech is enhanced by appropriate gesture, bumbling and hesitant speech in which the speaker has to disinter his thoughts from the dark catacombs of his mind, is often rendered farcical by stereotyped gestures which indicate an agony of mind rather than a fluency of mental imagery. Ethologists would refer to such movements as displacement activities.

Richness of gesture varies with race. D. Efrom and J. P. Foley studied gestures among Jewish and Italian immigrants to the United States and showed how the extroverted theatrical gestures of the Italians were in contrast to the more intellectual introverted hand movements of the Jews. Sorell regards the use of gesture as owing more to racial origins than to environmental influences. Desmond Morris describes three basic forms of hand waving: the *vertical*, the "*hidden palm*", wave, and the *lateral* wave. The "*hidden palm*" wave indicates that the palm is facing inwards while in the other two it is facing outwards. He describes the vertical wave as being common in France but rare or absent in Italy or Britain. The "hidden palm" wave is almost wholly an Italian gesture, and the lateral wave is predominantly British though not uncommon in France.

As a substitute for language difficulties, gestures are obligatory and are often combined with simple words like the immortal, "You Jane, me Tarzan" interchange. Modern sign languages evolved from the original system invented by Charles Michel in 1759 and taught to deaf mutes at a special school in Paris; nowadays a number of different versions of gestural speech exist, such as the Deaf and Dumb Language and the American Sign Language (ASL). Communication through signs was extensively practised by the Plains Indians to overcome the variety of languages and dialects among the nations. As far as I know there is no international sign language today comparable to the spoken lingua franca, pidgin or creole languages, though the ASL would probably do very well. Morris calls such "languages" coded gestures, a formal system of hand and arm signals. He includes semaphore and the tic-tac system used by bookmakers and their touts.

54. Hunting gestures used by Bushmen to transmit information regarding the nature of the game:
    (a) Ostrich
    (b) Eland
    (c) Bat-eared fox
    (d) Giraffe (Courtesy of Irvin DeVore, drawn from photographs).

Sign languages are also used when a ban on speech is enforced by circumstances – war parties, commando units and hunting groups. The Kalahari Bushmen have developed a sign language to communicate with their companions when stalking game; the gestures are pictorial rather than symbolic (Fig. 54), the hands mimicking the overt characteristics of the animal that has been discovered. Certain religious communities who have taken the vow of silence, such as the Trappist monks, employ a sign language as do those involuntary occupants of cells, the inmates of HM Prisons. The range of secret signs and handshakes falls into this category. Road-drillers, boiler-makers, shipyard workers, pop-groupers have, no doubt, their own ways of communicating non-verbally against a background of noise.

55. *Classical gestures used in acting and rhetoric in Elizabethan England (Courtesy of the Oxford University Press with Geoffrey Cumberlege, from* Elizabethan Acting *by B. L. Joseph).*

Cab and lorry drivers, too, have evolved their unofficial sign language. However, they include some gestures which will never be found in the Highway Code and which are both insulting and effective. The precise posture of the hand in executing a gesture is important and the distinctions are subtle. Consider the episcopal blessing, the Churchillian victory sign, and what has come to be known as the "Harvey Smith" gesture. They are all basically two-fingered, but the context in which they are expressed and the precise posture of the hand and forearm vary considerably. The first two-fingered gestures mentioned above are given with the palm facing outwards while an obscene version is performed with the palm facing inwards, often accompanied by an aggressive stabbing or jerking movement; furthermore, the obscenity of the gesture is increased by a reduction of the V-element, the smaller the angle subtended by the two fingers, the greater the insult. Churchill's fingers were always widely spread. His gesture bore no hint of vulgarity but was merely an exhortation to the people for further efforts and sacrifices in the common cause.

Gestures used in professional activities such as acting, mime, ballet and public speaking possess a ritualistic quality. The actions and gestures are linked to form set-pieces. In the words of John Bulwer, the author of *Chirologia* and *Chironomia* (two books originally published together in 1644):

> . . . the moving and significant extension of the *Hand* is knowne to be so absolutely pertinent to speech, that we together with a speech expect the due motion of the *Hand* to explaine, direct, enforce, apply, apparrell, and to beautifie the words men utter, which would prove naked, unless the cloathing Hands does neatly move to adorn and hide their nakedness, with their comely and minis-teriall parts of speech. (*Chironomia*, p. 16.)

Bulwer is concerned with rhetorical delivery but, as Joseph (1951) notes, in Elizabethan times what applied to gesture in oratory also applied to acting on the stage, although Bulwer takes care to distinguish between the miming of the stage player and rhetorical gestures of the orator that flow out of the "liquid current of nature" (Fig. 55).

> Gesture must attend upon every flexion of the voice, not scenical but declaring the sentence and the meaning of our minds, not by demonstration, but by signification . . . to represent a Physician feeling the pulse of the arteries . . . or to show a Lutenist striking the chords of an instrument are kinds of expression to be avoided . . .

To do so would have smacked of the dancer or actor; constantly he warned against gestures that, in their flamboyance and vehement delivery, belonged more to the theatre than to the forum. Joseph notes that in Cicero's *De Oratore* the same relation between acting and oratory must have existed in ancient Rome as in Elizabethan England.

If it is true that oratory is dead, then the murderers have been the cinematograph, radio and television – in that chronological order; the death blow being television. We are perhaps fortunate in Britain in having had at least two great orators in our life-time – Winston Churchill and David Lloyd George – cast in the classical mould. That at the turn of the century politicians were still striking "rhetorical" poses is attested by the evidence of the early cinematograph records. The coming of radio on the one hand led to a further decline in the graceful art of public speaking and gave way to mere tub-thumping in the market-place which Bulwer would have assigned in its flamboyance to the theatre and not to the "Churches, courts of commonpleas, and the Councell-Table". The true conventions of the debate in Britain are only to be found in the Union debating societies of Oxbridge.

The coming of television completed the process and in fact brought the art of rhetoric full circle. The modern politician or public figure is placed at a desk or in an armchair, depending on the degree of formality required, his arm and hand movements are thus constrained and most of the time are out of shot. He hasn't a chance to irritate the viewer by flinging his arms about and his face is frequently held in close-up so that extravagant facial expressions are also denied him if he doesn't want to appear comic or absurd. He is therefore left with the cadences and modulations of his voice and the more subtle expressions of his face to embellish his oratory. For this art a certain amount of training is required. For all Bulwer's precepts, the television orator has to learn to be an actor. From the elegance of rhetoric, the art of the gentleman and intellectual, we have turned full circle to the actor who, perhaps, started it all.

Gestures of Dancing

The history of the dance is an even more ancient one. There is some evidence, though speculative, that dancing was practised in the late Palaeolithic. The Franco–Cantabrian cave paintings show groups of figures in dance-like formations, an observation which suggests that the dance was an element in the religio-mystic ritual that underlie the cave-paintings themselves.

There is a startling difference in the anatomy of the traditional dances of the East and West. The East (India, Southeast Asia) has

56. Expressive gestures of a Balinese dancer.

specialised in somewhat languid dances in which footwork is subordinate to intricate movements of the upper part of the body, especially the head, neck and hands (Fig. 56). This sort of dance is practised by people who are slim, flexibly jointed and lightly boned, wearing a minimum of clothing (as befits the climate) that does not hinder free movement. It is reputed that Indian dancing comprises at least 4,000 positions of the hands and fingers which, when choreographed, tell a story or illustrate a myth. In contrast, the hand in Balinese dances does not tell a story specifically but is used to decorate the movements of the dance as jewels beautify the hand itself. On Samoa and many other Polynesian islands, dances that consist almost purely of hand gestures are characteristic.

The traditional dances of the West, on the other hand, are in every way opposite. They do not favour the upper part of the body; they are energetic, muscular and often fast and furious, demanding intricate foot-work, such as flamenco, African dances of all kinds, tarantellas and the wild dances of Southern Europe. In more placid vein and cooler countries clog-dances and Morris dances, including "tap" (derived from clog-dancing), tend to display activities that do not lead to undue heat loss.

Males usually play the positive role in Western folk-dancing, women in the East. Compared with the Mudras, the symbolic sign-language of Indian dancing, the mime gestures of Western dance including ballet are relatively crude and simplistic; death, life, love, marriage, etc., are stylised in manual gesture, elegant but unsubtle. Here again the Western preference for "footwork" over "handwork" is apparent.

Gestures in Apes

It is only among the great apes, the possessors of true hands, that gesture is commonly seen and, among them, it is only in the chimpanzee that it is so highly developed. A pleading or begging gesture with the hand, palm extended and facing upward, is often performed by macaques. I believe that it is not part of their repertoire in the wild but copied from the humans about them. Temple monkeys in India, provisioned and partly wild Japanese macaques and Gibraltar "apes" have been observed begging. In zoos, of course, all monkeys have learned this trick, including South American spider monkeys which "beg" by sticking their prehensile tails through the bars of the cage. The gorilla's use of the hand seems mainly to be tied up in ritualised displays like chest-beating, rather than for casual gestures, as is the chimpanzee's. The orang-utan is not known to make any specific gesture but, compared with the gorilla and chimpanzee, our knowledge of their naturalistic behaviour is rather poor. Chimpanzee gestures are very graphic and remarkably man-like. There is a dismissive gesture accompanied by a threatening face, a comforting gesture in which the infant or adolescent is gently patted by adult males, greeting gestures in which the hand is held out by the male to receive a kiss from the adult female, and patting-on-the-back and lip-kissing gestures when chimpanzees mutually embrace. Contact-seeking behaviour, touching and embracing bring about much reassurance and comfort. All these gestures and many more have been observed by Jane Goodall in her twenty-year study of wild chimpanzees in Tanzania.

The hallmarks of mankind are those characters which separate man from his primate relations. There are not many of them. The differences for the most part are quantitative, man being at one end of a spectrum of variation and apes and monkeys at the other. Some characteristics, like tool-making, enjoyed a brief reign as the unique criterion of humanness until Goodall's chimpanzees showed us that even chimps *make* tools (in a modified sort of way), so tool-making is no longer quite the acid test it was once considered. A critical brain size, known as "Keith's Rubicon" was used as the principal touchstone of humanness for many years. Speech and language is now the vogue. Human speech seems to be quite unique – out on its own. As Jane Lancaster put it a few years ago: "The more we know about communication (in apes and monkeys) the less these systems seem to help in our understanding of human language".

The essence of human speech lies in a number of unique factors that have been called "design features" by Charles Hockett. Essentially, such design features allow the communication of abstract ideas, the discussion of events displaced in space and time, the transmission of information through teaching and learning and the invention of new words. Human speech is not stereotyped like the closed systems of animals such as the chimpanzee, but is open and fluid, constantly ebbing and flowing with changing fashion and the stimulus of human events.

Chimpanzees, because of their overt intelligence as well as their

Gestures

friendly and co-operative temperaments, have often been the subjects of language and speech experiments. Initially, investigations were directed towards teaching chimpanzees to speak, though the projects were singularly unsuccessful. The logical follow-up was to investigate the use of gesture and mime. This has proved a much more successful venture, and communication, albeit at a simplistic level, was established.

One of the best-known experiments was carried out by Alan and Beatrice Gardner, a husband-and-wife team of psychologists. Washoe, their subject, was a young female adolescent when they started work (Fig. 57). The system they employed was the American Sign Language. The ASL is not a form of finger spelling, for each

57. *Washoe making the sign for toothbrush (Courtesy of R. A. and B. T. Gardner).*

position of the hand corresponds to a different word. The Gardners called them "signs"; others have referred to them as "symbols" or "words".

Washoe was taught signs in a number of different ways. One of the ways was to mould a similar and pre-existing natural gesture into an acceptable ASL sign. The chimpanzee's begging gesture with palm facing upward was refined into the gesture indicating "Give me". Its derivative, in which a beckoning gesture was added, became "Come". She was also taught by placing her hands in the appropriate positions and rewarding her. Finally she acquired some of her words through observational learning: "tooth-brush", for example. Washoe, like a young child, indulged in a certain amount of non-verbal "babbling", practising silently the words she had learned.

Occasionally she would invent a new "word", such as the famous episode of the bib for which the Gardners could find no sign in the ASL manual. They developed a sign for the word but Washoe had ideas of her own and *she* invented her own sign which in the event turned out to be the correct ASL-ese for bib.

By all these methods Washoe acquired a rich vocabulary of signs numbering nearly 100 in all. Her grammar was of the simplest and consisted of two-word phrases like "Washoe sorry", "Please tickle" and "Drink red". Washoe's achievements with two-word combinations are of a similar level to human children's earliest attempts to speak grammatically.

The Gardners tell a story about the meeting of Washoe and a young deaf child at a party. They looked at one another and then solved a socially awkward confrontation in their own unselfconscious way. Washoe rocked her arms, the sign for a baby, and the infant scratched at its armpit, the sign for a monkey.

The widespread use of non-verbal communication by means of manual gesture among human beings and the facility with which chimpanzees will learn the skill, in spite of their relatively small brain-size, strongly suggests that the combination of vocal calls and sign language would have served early hominids well enough until speech and language evolved. It has been said by J. S. Weiner, an authority on man in the Pleistocene, that "it is difficult to believe that co-operative hunting could be performed or skilled tool-making taught without language". Yet hunting, as discussed above, is a silent activity with its own sign language conventions. What might be more difficult without a symbolic language would be to plan the hunt ahead of time, but why should planning be so important? Each individual would know by experience his role in the hunting party and would behave accordingly.

As far as tool-making is concerned, demonstration is worth a thousand words and an observant apprentice to the tribal tool-maker would soon pick up the principals of the technique. Although human in every way – walking, tool-making, hunting – early man could still have operated effectively without a language. We really do not know at what stage speech (the anatomical component) and language (the grammatical) were established. It could have been at *Australopithecus* level, or it could have been as late as the earliest *sapiens* populations of the middle Pleistocene. If brain-size is any criterion, then one might favour the time of emergence of *Homo erectus* as the critical period. Between *Australopithecus* and *Homo erectus* levels the capacity of the brain almost doubled, from 508 cc to 974 cc.

Manual gestures are movements we all make as an integral part of our communication system. Sometimes they serve to replace speech, sometimes to augment and elaborate it and sometimes to render it farcical as did Hamlet's Players, and for which they were properly disciplined: ". . . do not saw the air too much with your hand, thus".

One way or another, gesture constitutes an enrichment of communication and a betterment of understanding between individuals. It is the hidden language that has no vocabulary and no grammar. It allows things to be expressed that can never be spoken of. If language was given to men to conceal their thoughts, then gesture's purpose was to disclose them.

58. Sketch of hands by Leonardo da Vinci.

Suggested Reading and References

A. General

Bandi, Hans-Georg 1961 *The Art of the Stone Age*. London: Methuen.

Barsley, Michael 1966 *The Left-handed Book*. London: Souvenir Press Ltd (also published by Pan Books, 1969).

Bell, Sir Charles 1834 *The Hand; its mechanism and vital endowments as evincing design*. London: William Pickering.

Brandon, James R., Clarke, Mary and Koegler, Horst 1975 Dance, The Art of (including Dance, Western and Dance, Eastern). *Encyclopaedia Brit*. 15th Edition.

Cherrill, Frederick R. 1954 *The Fingerprint System at Scotland Yard*. London: HMSO.

Goodall, Jane (see Van Lawick Goodall, Jane).

Joseph, B. L. 1951 *Elizabethan Acting*. Oxford University Press with Geoffrey Cumberlege.

Leakey, L. S. B. 1933 *Adam's Ancestors*. London: Methuen.

Morris, Desmond 1967 *The Naked Ape*. London: Jonathan Cape, Ltd.

Morris, Desmond 1977 *Manwatching*. London: Jonathan Cape, Ltd.

Napier, John 1962 The evolution of the hand. *Scientific American*, **207**, 6: 56–62.

Oakley, Kenneth P. 1967 *Man the Tool-maker*. London: Trustees of the British Museum (Natural History).

Sorell, Walter 1967 *The Story of the Human Hand*. London: Weidenfeld & Nicolson.

Van Lawick Goodall, Jane 1971 *In the Shadow of Man*. London: Collins.

Washburn, S. L. 1960 Tools and Human Evolution. *Scientific American*, **203**: 63–75.

Wile, Ira S. 1934 *Handedness: Right and Left*. Boston: Lothrop, Lee & Shepard Co.

B. Specialist

Barron, J. N. 1970 The structure and function of the skin of the hand. *British Journal of Surgery of the Hand*, **2**: 93–99.

Brooker, D. R. 1977 Henry Faulds (1843–1930). *Fingerprint Whorld*, **2**: 59–64.

Bunnell, Sterling 1944 *Surgery of the Hand*. Philadelphia, USA: J. B. Lippincott Co.

Capener, Norman 1952 Leonardo's Left Hand. *The Lancet*, 19 April: 813–816.

Clark, W. E. Le Gros 1936 The problem of the claw in primates. *Proc. zool. Soc. Lond*: 1–24.

Clarkson, Patrick and Pelly, Anthony 1962 *The General and Plastic Surgery of the Hand*. Oxford: Blackwell.

Cummins, Harold 1941 Ancient finger prints in clay. *Scientific Monthly*, **7**, 5: 389–402.

Dart, R. A. 1957 *The Osteodontokeratic Culture of Australopithecus prometheus*. Pretoria: Transvaal Museum Memoirs, No. 10.

Day, Michael H. 1977 *A Guide to Fossil Man*, 3rd Edition. London: Cassell.

Davis, D. Dwight 1964 *The Giant Panda. Fieldiana: Zoological Memoirs No. 3*.

Efrom, D. and Foley, J. P. 1941 *Gesture and Environment*. New York: King's Crown Press.

Goodall, Jane (see Van Lawick Goodall, Jane).

Grand, T. I. 1972 *J of Mammalogy*, **53**, 1: 198–201.

Harrison, Richard J. and Montagna, William 1969 *Man*. New York: Appleton-Century-Crofts.

Holt, Sarah B. 1968 *Genetics of Dermal Ridges*. Springfield, Ill: Charles C. Thomas.

Jones, F. Wood 1916 *Arboreal Man*. London: Edward Arnold.

Jones, F. Wood 1942 *Principles of Anatomy as seen in the Hand*, 2nd Edition. London: Bailliere, Tindall & Cox.

Kaplan, E. B. 1953 *Functional and Surgical Anatomy of the Hand*. Philadelphia, USA: J. B. Lippincott Co., London: Pitman Medical Publishing Co., Ltd.

Kidd, Walter 1907 *The sense of touch in mammals and birds*. London: Adam and Charles Black.

Lever, J. D.   1963   *The Hand of Man*. Cardiff: University of
  Wales Press.

Musgrave, Jonathan   1970   How dextrous was Neanderthal
  Man? *Nature*, **233**: 538–541.

Napier, J. R.   1956   The prehensile movements of the human
  hand. *J. Bone Jt. Surg.* **38B**: 902–903.

Napier, J. R.   1960   Studies of the hands of living primates.
  Proc. zool. Soc. Lond. **134**: 647–657.

Napier, J. R.   1961   Prehensility and opposability in the hands
  of primates. *Symp. zool. Soc. Lond.* **10**: 115–132.

Napier, J. R.   1962   Fossil hand bones from Olduvai Gorge.
  Nature, **196**: No. 4853: 409–411.

Napier, J. R.   1976   *The Human Hand*. Carolina Biological
  Readers, No. 61. Carolina Biological Supply Co.

Napier, J. R. and Davis, P. R.   1959   *The Forelimb Skeleton and
  Associated Remains of Proconsul africanus. Fossil Mammals of
  Africa*, No. 17. London: Trustees of the British Museum
  (Natural History).

Oakley, Kenneth P.   1958   Tools makyth man. *Smithsonian
  Report for 1958*: 431–455.

Tuttle, Russell H.   1967   Knuckle-walking and the evolution of
  hominoid hands. *Am J. phys. Anthropology*, **26** (2): 171–206.

Van Lawick Goodall, Jane   1968   Behaviour of Free-living
  Chimpanzees in the Gombe Stream Reserve. *J. Animal
  Behaviour*, **1**, 3: 161–311.

Index

Index

Index

9203   01 446762   01 5   (IC=2)
NAPIER. JOHN RUSSELL   02/27/84
HANDS
(4) C1980 1ST AMERIC MAIN QM54

QM548.N35 1980
Hands

3 2551 000089866

QM   HANDS
548
.N35   087075   NAPIER
1980

NORMAN F. BOURKE
MEMORIAL LIBRARY
CAYUGA COMMUNITY
COLLEGE
AUBURN, N. Y. 13021